Jens Müller

Characterisation of local aluminum-alloyed contacts

Jens Müller

Characterisation of local aluminum-alloyed contacts
to silicon solar cells

Südwestdeutscher Verlag für Hochschulschriften

Impressum / Imprint
Bibliografische Information der Deutschen Nationalbibliothek: Die Deutsche Nationalbibliothek verzeichnet diese Publikation in der Deutschen Nationalbibliografie; detaillierte bibliografische Daten sind im Internet über http://dnb.d-nb.de abrufbar.
Alle in diesem Buch genannten Marken und Produktnamen unterliegen warenzeichen-, marken- oder patentrechtlichem Schutz bzw. sind Warenzeichen oder eingetragene Warenzeichen der jeweiligen Inhaber. Die Wiedergabe von Marken, Produktnamen, Gebrauchsnamen, Handelsnamen, Warenbezeichnungen u.s.w. in diesem Werk berechtigt auch ohne besondere Kennzeichnung nicht zu der Annahme, dass solche Namen im Sinne der Warenzeichen- und Markenschutzgesetzgebung als frei zu betrachten wären und daher von jedermann benutzt werden dürften.

Bibliographic information published by the Deutsche Nationalbibliothek: The Deutsche Nationalbibliothek lists this publication in the Deutsche Nationalbibliografie; detailed bibliographic data are available in the Internet at http://dnb.d-nb.de.
Any brand names and product names mentioned in this book are subject to trademark, brand or patent protection and are trademarks or registered trademarks of their respective holders. The use of brand names, product names, common names, trade names, product descriptions etc. even without a particular marking in this works is in no way to be construed to mean that such names may be regarded as unrestricted in respect of trademark and brand protection legislation and could thus be used by anyone.

Coverbild / Cover image: www.ingimage.com

Verlag / Publisher:
Südwestdeutscher Verlag für Hochschulschriften
ist ein Imprint der / is a trademark of
OmniScriptum GmbH & Co. KG
Heinrich-Böcking-Str. 6-8, 66121 Saarbrücken, Deutschland / Germany
Email: info@svh-verlag.de

Herstellung: siehe letzte Seite /
Printed at: see last page
ISBN: 978-3-8381-3833-6

Zugl. / Approved by: Hannover, U, Diss., 2013

Copyright © 2014 OmniScriptum GmbH & Co. KG
Alle Rechte vorbehalten. / All rights reserved. Saarbrücken 2014

Kurzzusammenfassung

Wir untersuchen in dieser Arbeit die Rekombination an lokalen Aluminium legierten Kontakten, die durch lokale Laserablation einer Schicht von Dielektrika und anschließendem ganzflächigem Druck von Aluminium Paste hergestellt werden. Die Kontaktbildung während eines abschließenden Temperschritts wurde erstmalig quantitativ beschrieben. Der Schwerpunkt der Arbeit ist die Analyse der Herstellungsparameter von Al legierten Kontakten um möglichst niedrige Kontaktrekombination in der Solarzelle zu erreichen.

Wir beschreiben die lokale Kontaktrekombination mit einem analytischen Modell. Zusammen mit ortsaufgelösten Lebensdauermessungen mittels der dynamischen ILM Methode unterscheiden wir zwischen der Rekombination im passivierten und kontaktierten Gebiet. Wir testen diese Methode an lokalen Kontakten, die durch Laserablation einer Schicht von Dielektrika und anschließendem Aufdampfen von Al hergestellt wurden (LCO). Laser Fired Contacts (LFC) wiesen ähnliche Rekombinationeigenschaften wie LCOs auf.

An lokalen Al legierten Kontakten messen wir Sperrsättigungsstromdichten $J_{0,cont}$ bis zu 9×10^2 fA/cm^2 was mindestens eine Größenordnung unter den Werten für LCO und LFC liegt. Die Kontaktrekombination wird durch eine hoch Al dotierte (Al-p^+) Schicht der Dicke $W_{Al-p^+} > 1\mu m$ verringert. Durch Analyse des Kontaktbildungsprozesses als Funktion der Kontaktgeometrie zeigen wir, dass Punktradien $r > 100$ μm und Linienbreiten $a > 80$ μm notwendig für minimale Rekombination sind.

Mit Hilfe analytischer Modellbildung berechnen wir W_{Al-p^+} in Abhängigkeit der Prozessparameter. Hierfür beschreiben wir die Zeitabhängigkeit der Siliziumkonzentration c_{Si} in der Aluminiumschmelze mit einer Differentialgleichung erster Ordnung. So können wir W_{Al-p^+} in Übereinstimmung mit experimentellen Ergebnissen berechnen. Als Resultat der Prozessdynamik finden wir das c_{Si} kleiner als die Gleichgewichtskonzentration im Phasendiagramm sein kann. Wir berücksichtigen in unserem Modell den Einfluß des Kontaktabstandes auf W_{Al-p^+}.

Ein Vergleich von ns und ps Laserablationsprozess zeigte um eine Größenordnung höhere J_0 Werte für den ns Prozess. Die starke Inhomogenität der Al-p^+ Schicht wird durch eine erhöhte Rauhigkeit der Oberfläche nach der ns Laserablation verursacht. Durch einen kurzen KOH Ätzschritt nach der ns Laserablation ist es gelungen die Oberflächenrauhigkeit so zu verringern, dass ns und ps Laserkontakte ähnliche Eigenschaften in Bezug auf Rekombination und Kontaktbildung zeigen.

Wir schätzen das Potential lokaler Al legierter Kontakte mit Simulationen von Solarzellenparametern ab. Dafür erweitern wir ein Modell für die Optimierung der Rückseitenkontaktgeometrie um experimentell verifizierte Parametrisierungen der Rekombination, des Serienwiderstandes und der Rückseitenreflektion von lokalen Al legierten Kontakten. Unsere Simulationen zeigen, dass Punkt- und Linienkontakte gleiche Wirkungsgrade erzielen bei einem optimalen Metallisierungsgrad von 20% auf 2 Ωcm p-Typ Czochralski Silizium.

Stichworte: Siliziumsolarzelle, lokale Al Legierung, Rekombination, kinetisches Modell, Laserablation, analytische Simulation

Abstract

We analyze in this work the recombination at local aluminum alloyed contacts, which are realized by local laser ablation of a dielectric stack and subsequent full area screen printing of Al paste. The contact formation occuring during a final rapid thermal anneal is quantitatively described for the first time. We focus in this work to analyze the processing parameters of local Al alloyed contacts to enable lowest contact recombination in solar cells.

We study the local contact recombination employing an analytical model. Together with spatially resolved lifetime measurements obtained from dynamich Infrared Lifetime Mappings (ILM), we separate recombination in the passivated and contacted regions. We test this technique at local contacts formed by laser ablation of a dielectric stack and subsequent evaporation of Al (LCO). For laser fired contacts (LFC) we find equal contact recombination.

At local Al alloyed contacts we determine contact reverse saturation current densities as low as $J_{0,cont} = 9 \times 10^2 \; fA/cm^2$. This value is at least one order of magnitude lower compared to those determined at LCO and LFC. From scanning electron microscopy (SEM) images we find reduced recombination as a result of more than 1 μm thick highly Al doped (Al-p^+) layers. Analyzing the contact formation process as a function of the contact size and layout we show that point contact radii $r > 100 \; \mu m$ and line contact widths $a > 80 \; \mu m$ are appropriate for lowest contact recombination.

Using quantitative analytical modeling we describe the Al-p^+ layer thickness W_{Al-p^+} as a function of the processing parameters. With a first order differential equation we describe the time dependence of the silicon concentration c_{Si} in the aluminum melt. As a result we are able to predict W_{Al-p^+} in accordance with measured values. We find that c_{Si} is smaller than the equilibrium Si concentration as a result of the process dynamics such as the dissolution rate of solid silicon and the transport of silicon in the aluminum melt. We implement the effect of contact spacing on W_{Al-p^+} in our model.

We compare ps and ns laser ablation processes used to create local Al alloyed contacts. We observe one order of magnitude higer J_0 values at the contacts processed with ns laser pulses. A strong inhomogeniety of the contact formation process and thus W_{Al-p^+} were interpreted as a result of the strong surface roughness after ns laser ablation. Introducing a short KOH etch after laser ablation reduces surface roughness and results in compareable properties of ns and ps laser process in terms of recombination and contact formation.

To indentify the potential of local Al alloyed contacts in silicon solar cells we adopt their analysis to simulations. For this purpose we extend an optimization tool for the rear contact geometry of solar cells with experimentally verified parameterizations of the recombination, series resistance and the rear contact fraction dependent rear reflectance. Our study reveals equal performance of point and line rear contact layouts with an optimum metallization fraction of 20 % on 2 Ωcm p-type Czochralski grown silicon.

Keywords: silicon solar cell, local aluminum alloying, recombination, kinetic model, laser ablation, analytic simulation

Contents

Introduction   7

1. Local rear contacts   11
    1.1. Advantages 11
    1.2. Local contact types 12
    1.3. Aluminum alloyed contacts 14
        1.3.1. The alloying process 14
        1.3.2. Structure of local Al alloyed contacts 16

2. Measuring recombination at local contacts   19
    2.1. Recombination current and lifetime measurements 19
    2.2. Dynamic Infrared Lifetime Mapping (ILM) 21
        2.2.1. Setup 21
        2.2.2. Principle 22
        2.2.3. Calibration of static ILM images 23
    2.3. Separation of recombination in the passivated and contacted area 24
        2.3.1. Derivation of the Fischer model 25
        2.3.2. The effective rear surface recombination velocity 26
        2.3.3. Restrictions for the determination of S_{cont} 28
    2.4. Application example 29
        2.4.1. Sample preparation 29
        2.4.2. Effective rear surface recombination 31

	2.4.3. Contact recombination velocity	32
	2.4.4. Impact of the doping density	35
2.5.	Conclusion	37

3. Recombination at local Al alloyed contacts 39

3.1.	Sample preparation	39
3.2.	Contact recombination	40
	3.2.1. Effective lifetime measurements	40
	3.2.2. Effective rear surface recombination velocity	41
	3.2.3. Contact recombination velocity	43
3.3.	Structural investigation	46
	3.3.1. Al-p^+ layer thickness	46
	3.3.2. Predicting the contact recombination velocity	47
3.4.	Conclusion	50

4. Understanding the local alloying of aluminum and silicon 53

4.1.	Kinetic model of the local alloying process	54
	4.1.1. The firing step	54
	4.1.2. Silicon concentration in time	55
	4.1.3. Al paste consumption	56
	4.1.4. Dissolution of silicon in the melt	57
	4.1.5. Experimental verification of the model	59
4.2.	Considering spacing of the contacts	61
	4.2.1. Extension of the kinetic model	62
	4.2.2. Impact of peak firing temperature	64
4.3.	Predictions of this model	64
4.4.	Conclusion	67

5. Ablation of the dielectric layer 69

5.1.	Comparison of ps and ns laser processes	69

		5.1.1. Contact recombination	70
		5.1.2. Al-p^+ layer thickness	71
	5.2.	Introducing an additional KOH etch	73
		5.2.1. Process optimization	73
		5.2.2. Comparison of ps and ns laser process with subsequent KOH etch	75
	5.3.	Conclusion	77

6. Simulation of silicon solar cells with local Aluminum alloyed base contacts 79
    6.1. Device Model 79
        6.1.1. Contact recombination 81
        6.1.2. Base series resistance 83
        6.1.3. Optics 83
    6.2. Simulation results 86
        6.2.1. Comparison with literature cell data 87
        6.2.2. Point or line contacts? 89
        6.2.3. Comparison of local Al alloyed contacts opened with the ps and ns laser 90
        6.2.4. Impact of contact resistance 90
    6.3. Conclusion 91

7. Summary 93

A. Comparison of Fischer model and numerical simulations 95

B. Other cell simulation results 99
    B.1. Impact of rear reflectance 99

References 103

List of publications 113

Introduction

A silicon solar cell converts the power of the incident light directly into electrical power. The process of power conversion can be divided into three basic mechanisms:

1. **Absorption** of the incident light in the semiconducting material.
2. **Separation** of electrons and holes in an electrical field.
3. **Transport** to the terminal contacts to provide the electrical energy to the load.

Improving any of those three mechanisms may result in an increase of the efficiency of the energy conversion process. By reducing the contact resistance of the solar cell for example it is possible to use a higher amount of the power generated within the cell. However, when applying electrical contacts to solar cells they may as well influence the other two mechanisms and create therefore a complex interdependency.

In particular is the separation of electrons and holes significantly reduced due to a high degree of recombination which is commonly observed at metal contacts. To reduce the contact recombination a highly Al doped (Al-p^+) layer was introduced below the solar cell base contact as early as in the 1970's [1]. In mass production of silicon solar cells this layer is fabricated by alloying of Al and Si at the entire rear surface [2]. Another concept of reducing contact recombination has been introduced with the point contacted back junction solar cell [3] and was later succesfully applied to front junction solar cells [4,5]. Here the base material has been only locally contacted reducing the area where recombination can occur.

In this work we measure the recombination at local contacts to silicon solar cells. Of special interest is hereby the combination of the local rear contact concept and alloying of Al and Si as it is a promising way to increase the efficiency of industrially fabricated silicon solar cells [6–10].

During alloying of Aluminum and Silicon a highly aluminum doped (Al-p^+) layer is formed [11]. This layer is known to reduce contact recombination [12]. However, from earlier experiments [13–15] it is known that the formation of local Al alloyed contacts strongly depends on the processing conditions. As a result it remained unclear how to create a sufficiently thick local Al-p^+ layer and more importantly how to achieve low contact recombination with this concept.

We introduce a technique to determine the local contact recombination based on lifetime measurements employing the dynamic Infrared Lifetime Mapping (dynamic ILM) method [16]. For this purpose we use an analytical model to separate recombination in the passivated and contacted region [17] We apply this technique to local Al alloyed contacts analyzing the basic mechanism of reduced contact recombination. Furthermore we develop a general understanding of the local contact alloying as a function of the processing conditions. For this purpose we investigate how ablation of the dielectric layer influences contact recombination. Finally, we evaluate our findings regarding the application of local Al alloyed contacts to solar cells.

The outline of this work is as follows:

Chapter 1 gives an introduction into the concept of local contacts. Furthermore we explain the theory of contact recombination and point out two ways to reduce it. Finally we review the concept of local Al alloyed contacts with a brief summary of Si and Al alloying, a description of the manufacturing process and the contact structure.

Chapter 2 deals with the technique to quantify the local contact recombination. First we analyze the link between recombination properties of solar cells and quantities determined from lifetime measurements. Then we introduce the dynamic ILM technique which is used for lifetime measurements at locally contacted samples. For the interpretation of the lifetime measurements in terms of local contact recombination we employ an analytical model. As this model is often misinterpreted we briefly discuss the basic principle and possible restrictions of its application. Furthermore we compare a numerical simulation with the model to justify its application. Finally we apply our technique for the determination of the local contact recombination to laser processed point contacts formed by Laser Contact Openings (LCOs) of a dielectric stack prior to the metallization [18] or after the metallization as Laser Fired Contacts (LFC) [19]. We investigate the recombination at those contacts and compare our results with literature values.

Chapter 3 describes the recombination at local Al alloyed contacts as a function of the local contact geometry, i.e. point or line geometries of different size. The highly aluminum-doped regions are analyzed in terms of Al-p^+ layer thickness by scanning electron microscope investigations. The results were correlated with the recombination properties and compared with our theoretical expectation.

Chapter 4 investigates how local Al alloyed contacts form. We derive a basic analytical model to describe the thickness of the Al-p^+ layer as a function of the processing conditions. For this purpose we consider the kinetics of the local alloying process allowing to determine a dissolution velocity of Si during alloying and the volume local alloying is restricted to. We extend this model to consider the impact of contact spacing on contact formation and experimentally verify our model. Finally, we derive a more general interpretation of the kinetic model.

Chapter 5 describes how the contact recombination is influenced by the laser process used for the ablation of the dielectric layer. For this purpose we compare a ps and ns Laser process. From this investigation we developed the idea to combine the ns laser process with a subsequent etch of the surface in KOH to reduce surface roughness. We optimize this process and compare it to a ps laser process in terms of contact recombination.

Chapter 6 introduces an analyitcal simulation of silicon solar cells from Wolf et al. [20]. With the help of this simulation we evaluate the properties of local Al alloyed contacts with respect to the application for solar cells. We extend the model from Wolf to consider the outstanding properties of local Al alloyed contacts and compare our simulation results with experimental data. Finally we examine the impact of contact geometry, the laser process used for ablation of the dielectric layer, rear reflectance and the specific contact resistance on solar cell performance.

Chapter 7 summarizes this work.

1. Local rear contacts

In this chapter we introduce the concept of local rear contacts. In Fig. 1.1 a front junction solar cell with a) full area and b) local rear contacts is depicted. The main difference of the two cell types is the introduction of a dielectric layer at the rear. This dielectric layer is interrupted by the local rear contacts.

Local rear contacts were first applied to back junction solar cells allowing for contacts of both polarities at the rear of the solar cell [3]. Avoiding contact shading on the front side, these cells have excellent optical properties and demonstrate very high generation currents. To front junction solar cells local rear contacts were introduced with the Passivated Emitter and Rear Cell (PERC) [4]. This cell type had the same structure as in Fig. 1.1b.

1.1. Advantages

With local rear contacts we aim at an increase of the power conversion efficiency η of the solar cell. Applying local rear contacts η is significantly improved by two mechanisms:

- Increased rear reflectivity

  Metal at the solar cell rear acts as a mirror. However, in the case of a dielectric layer between metal and Si the reflectance can be further increased. As a result we achieve a better light trapping in the solar cell. In this case the amount of absorbed light is increased which increases the total current that can be generated within the solar cell and therefore also the short circuit current density J_{sc}.

- Reduced recombination

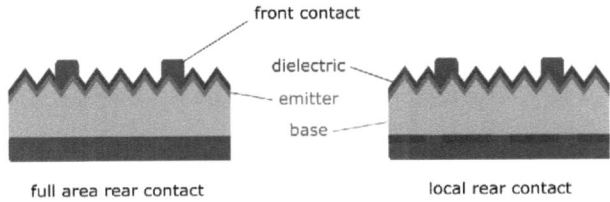

Figure 1.1.: *Comparison of a solar cell with a) full area and b) local rear contacts.*

At metal semiconductor interfaces the recombination of electron and hole pairs is large. Due to the introduction of a dielectric layer the area of the metal contacts is decreased. The reduced contact area as well as the low recombination at the dielectric layer lead to a significant reduction of the total recombination current, which improves the open circuit voltage V_{oc}.

However, majority charge carriers in the base have to travel larger distances in the case of a locally contacted solar cell, compared to a solar cell with full area rear contact. This transport issue might increase the solar cell series resistance and therefore decrease the fill factor FF. As a result also the power conversion efficiency

$$\eta = \frac{J_{sc}V_{oc}FF}{P_{light}} \tag{1.1}$$

of the cell will decrease. Here P_{light} denotes the incident light power.

We conclude from the arguments above, that the introduction of local contacts to the solar cell rear is non trivial. The application of local contacts may increase J_{sc} and V_{oc} but may lead to resistive losses which decrease the FF. The optimization of solar cells comprising local rear contacts is therefore always a trade-off between recombination and resistive losses. For this purpose several optimization studies of the contact geometry have been performed in the past [21–25].

1.2. Local contact types

The interface of Si to metal is highly recombination active. The recombination rate at metal contacts is only limited by the thermal velocity of minority charge carriers $v_{th} = 10^7 cm/s$ in the case of Silicon [26]. For this reason metal contacts may lead to substantial power losses in a solar cell even in the case of local contacts. For further optimization of locally contacted solar cells it is therefore desirable to reduce the recombination at metal contacts.

From the theory of surface recombination two ways are possible to reduce contact recombination [27]:

- Reducing the density of interface states

    The surface of the silicon crystal terminates the periodicy of the crystal lattice that is the basis of the formation of bonding and antibonding energy bands. Therefore, localized electronic states exist at the surface, which form recombination-active energy levels within the bandgap [28].

    To reduce the density of those interface states a dielectric layer between Si and metal is introduced [29]. However an electrical contact between Silicon and metal has to be maintained. For this reason the thickness of the dielectric layer is chosen sufficiently

1.2. Local contact types

thin to allow for tunneling of charge carriers. This technique has been applied to several dielectrics such as SiO_2 [29], SiN_x [30] as well as Al_2O_3 [31]. Even more effective is the use of amorphous Si in heterojunction solar cells allowing for very high open circuit voltages [32].

- Reducing the charge charrier density near the interface

    Both charge carrier types, i.e. electrons and holes, are necessary for recombination. Hence by decreasing the density of the minority charge carriers it is possible to reduce contact recombination effectively even in the case of a high density of recombination active states at the interface of Silicon and metal.

    To reduce the concentration of minority charge carriers usually a potential barrier at the interface is introduced [1]. This potential barrier can be generated applying a high doping concentration of majority charge carriers at the surface [33]. Employing this concept in combination with a local diffusion process resulted in the highest ever reported efficiency of a silicon solar cell [5].

In this work we focus on a reduction of contact recombination at locally rear contacted solar cells due to a reduced minority charge carrier concentration. For this purpose we apply a highly Aluminum doped (Al-p^+) layer generated by local alloying of Al and Si. A similar process is already effectively used for the mass production of Silicon solar cells while applying full area Aluminum alloyed contacts [1] to the solar cell rear.

For comparison we investigate in total three different local contact types in this work. They are possible candidates for mass production of solar cells due to their relatively low process cost:

- **Local Contact Opening (LCO)** - Contact is made by laser ablation of the dielectric layer and subsequent evaporation of Al [18].

- **Laser Fired Contacts (LFC)** - First Al is evaporated and then Si is contacted through the dielectric layer by laser [19]. For this type of contact the metal is fired through the dielectric layer by laser generating a local highly doped layer [34, 35].

- **Local Al alloyed contacts** - First LCO's are applied to the dielectric rear without subsequent evaporation of Al. Subsequently Al paste is screen printed on the rear and a high temperature step is applied. During the high temperature step a highly aluminum doped (Al-p^+) layer forms by local alloying of Al and Si.

Note, that also other concepts promise low contact recombination and low production cost. For instance laser processing has recently been intensively used to create a local highly doped layer employing several techniques [36, 37]. For this purpose we study the recombination at Laser Fired Contacts [19] in chapter 2 in more detail.

1.3. Aluminum alloyed contacts

The alloying of Al and Si is a process of enormous practical relevance especially in the photovoltaic industry. Applied to solar cells it forms a highly aluminum doped (Al-p^+) layer just beneath the electrical contacts [11]. The Al-p^+ layer features excellent electronic properties which even allows to use them as high quality emitters in solar cells [38, 39].

In the previous section we discussed how a highly doped layer at contacts to solar cells reduces charge carrier recombination. Therefore local Al alloyed contacts are of special interest for the application in locally contacted solar cells [6–10].

Local Al alloyed contacts are prepared using the following process sequence [40, 41]:

- Deposition of a dielectric layer stack.
- Local opening of the dielectric layer.
- Full area or local screen printing of an Al paste.
- Contact firing during a rapid thermal anneal.

Note, that a relatively smooth surface has to be provided for the formation of the local Al alloyed contacts with surface roughnesses at the order of $1\mu m$ [42, 43].

Optimizing the formation of local Al-p^+ layers for rear contacts through a locally opened dielectric layer, impressive efficiencies of up to $\eta = 21\%$ have been achieved with solar cells ready for mass production [10]. However several surprising phenomena were investigated, which were not fully understood. For example depends the thickness of the Al-p^+ layer strongly on the geometrical arrangements [44–46], the size [14, 15, 47] and spacing [48] of the contacts. Furthermore a strong impact of the Si content in the Al paste [7, 48] on contact formation has been observed. Moreover it was demonstrated that the Al-p^+ layer thickness and homogeneity strongly depends on the opening technique of the dielectric layer [44, 49].

As the Al-p^+ layer is essential for reduced contact recombination we aim at an improved understanding of these effects. But first we give here a short introduction into the concept of local Al alloyed contacts to create a basic understanding of this type of local contact. For this purpose we start with explaining briefly the alloying process of Al and Si.

1.3.1. The alloying process

The alloying of Al and Si takes place during a short firing step, the so-called rapid thermal anneal [40]. During this firing step the sample temperature increases to more than $800°C$. When the temperature reaches $660°C$ (the melting point of Al) the Al paste on top of the silicon sample becomes liquid. According to the phase diagram of Al and Si [50] shown in

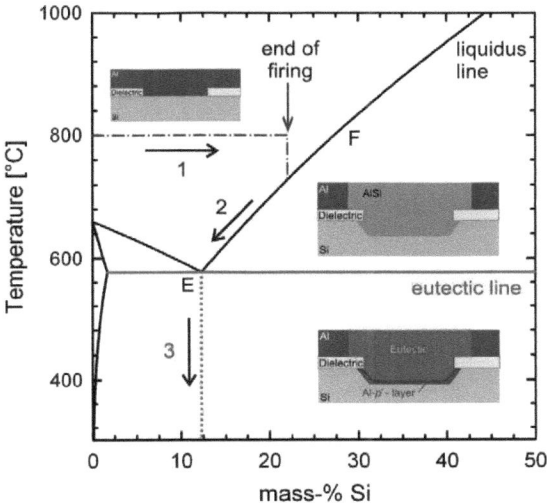

Figure 1.2.: *Equilibriium phase diagram of Al and Si after Ref. [50]. During firing (step 1) Si dissolves into the Al melt in the area of local contact openings (LCOs) as depicted in the inset. During cooling (step 2) liquid Si precipitates and the Al-p^+ layer is generated. After the eutectic temperature of 577°C is reached the remaining liquid solidifies with eutectic composition (step 3).*

Fig. 1.2 solid Si now dissolves into the Al melt although the melting point of Si at 1410°C has not been reached (step 1).

Note, in the case of local Al alloyed contacts a dielectric layer separates Al and Si, as demonstrated in the insets of Fig. 1.2. Here no alloying of Al and Si occurs unless the dielectric layer remains stable during firing [51,52]. However, in the areas of local contact openings (LCOs) a direct interface of Al and Si exists. Only in these areas alloying of Al and Si can happen.

The concentration of Si c_{Si} in the Al melt increases, since more and more solid Si dissolves into the Al melt. Theoretically the Si concentration c_{Si} saturates when the liquidus line is reached in Fig. 1.2. The firing step however may be finished before the liquidus line is reached. In this case the Si concentration does not increase further at the end of the firing.

After firing the sample cools down and liquid Si precipitates due to a limitation of its solubility in the alloy (step 2). The expelled Si atoms grow epitaxially onto the Si substrate and Al is incorporated into the lattice according to its solid solubility in Si forming the

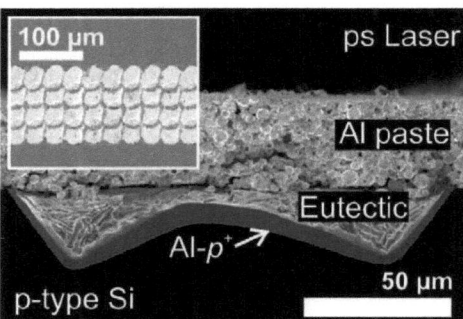

Figure 1.3.: *SEM image of a pico-second laser line contact. The Al-p^+ region appears brighter than the high resistivity bulk. (Upper left) Micrograph of the contact before screen printing. The dielectric layer appears blue, whereas the laser openings appear white in this example.*

Al-p^+ layer. When the liquid phase reaches the eutectic composition of $E \sim 12\%$ no more Si segregates and the remaining liquid phase solidifies (step 3).

The doping concentration in the Al-p^+ layer is orders of magnitude higher compared to doping concentrations usually used for Si solar cell base material [1]. As a result a potential barrier for minority charge carriers is induced [33] effectively preventing recombination at the Al alloyed contact.

1.3.2. Structure of local Al alloyed contacts

We introduce the typical structure of local Al alloyed contacts in Fig. 1.3. In the scanning electron microscope (SEM) image with secondary electron contrast the Al-p^+ layer appears brighter than the high resistivity bulk. On top of the Al-p^+ layer we find the eutectic layer with a Si concentration of $E = 12\%$. Above we observe the Al paste. In chapter 3 we will use SEM images of local Al alloyed contacts to measure the Al-p^+ layer thickness W_{Al-p^+} and correlate it to the measured contact recombination.

The dielectric layer in Fig. 1.3 has been opened with a ps laser. However also different laser processes or other techniques have been used to generate the local contact opening (LCO) [7, 44, 49, 53]. The impact of the laser type on contact formation will be discussed in chapter 5.

The micrograph in the upper left of Fig. 1.3 shows the LCO before screen printing. It reveals a line geometry of the contact. Two local contact geometries have been analysed in the literature so far: a) parallel line and b) equally spaced point contacts.

For comparison we show a point contact with similar dimensions in Fig. 1.4. As the

1.3. Aluminum alloyed contacts

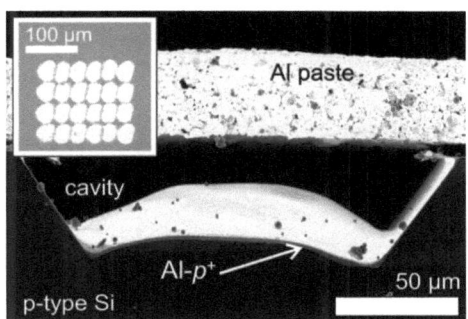

Figure 1.4.: *SEM image of a point contact with similar dimensions compared to the line contact in Fig. 1.3. However, instead of an eutectic layer a cavity is formed and the Al-p^+ layer thickness is significantly smaller compared to Fig 1.3.*

most striking difference between Fig. 1.3 and Fig. 1.4 we find a large cavity instead of the eutectic layer. This phenomena has been often discussed in the literature [13, 54–57]. So far no detrimental effects of cavities for solar cell operation have been reported, however avoiding cavities is found to result in a higher efficiency η of the device [56]. Furthermore is the Al-p^+ layer found to be considerably thinner compared to Fig. 1.4. We analyse the impact of the contact geometry on the contact formation in more detail in chapter 3.

2. Measuring recombination at local contacts

In this chapter we introduce our measurement technique to quantify recombination at local contacts represented by the contact recombination velocity S_{cont}. We measure the recombination current using lifetime measurements on specifically prepared test samples. This allows us to separate the recombination at the sample rear from other sources of recombination, i.e. bulk or front surface recombination.

For the effective charge carrier lifetime measurement we employ the dynamic Infrared Lifetime Mapping (ILM) technique. Introduced by Ramspeck et al. this calibration-free method allows to determine the spatially resolved effective lifetime of metallized samples [16]. Using an analytical model [17] we are able to distinguish between recombination at local contacts and the passivated interface in between the contacts. We examine this technique at common evaporated contacts before we apply it to local Al alloyed contacts. We published this work already elsewhere [58].

Before we start with the details of the measurement, we shortly review some basic principles of charge carrier recombination. This will provide a clear link between the recombination current at an interface in a solar cell and the lifetime measurements we perform.

2.1. Recombination current and lifetime measurements

The recombination current at an interface i in a solar cell is expressed by

$$J_{rec} = J_{0,i}\left[exp\left(V_i/V_{th}\right) - 1\right]. \tag{2.1}$$

Here V_i is the local voltage and V_{th} the thermal voltage. The reverse saturation current density $J_{0,i}$ is hence a measure for the recombination of charge carriers at this interface. Neglecting resistive losses or space charge recombination it directly translates into a current loss in the IV-characteristic

$$J_{rec} = J_{0,i}\left[exp\left(V_i/V_{th}\right) - 1\right] - J_{sc} \tag{2.2}$$

of a solar cell, where J_{sc} is the short circuit current density.

To separate recombination at the several interfaces of solar cells specific test samples are used. They allow to determine $J_{0,i}$ by measuring the effective charge carrier lifetime at only one interface i at a time. This approach is also beneficial in terms of process optmization, since it is not necessary to fabricate a complete device to determine $J_{0,i}$ of the single interface i.

Measuring the effective lifetime τ_{eff} delivers the total recombination rate

$$\frac{1}{\tau_{eff}} = \frac{1}{\tau_{bulk}} + \frac{1}{\tau_{surface}} \qquad (2.3)$$

in the sample. It is the sum of the recombination rates in the volume and at the surfaces of the device. Here τ_{bulk} equals the bulk lifetime. The surface recombination rate $1/\tau_{surface}$ can be approximated by [59]

$$\frac{1}{\tau_{surface}} = \frac{S_{front} + S_{rear}}{W}. \qquad (2.4)$$

Here S_{front} denotes the front and S_{rear} the rear surface recombination velocity. Eqn. 2.4 holds as long as the surface recombination is not limited by the transport of charge carriers to the surface, which is the case for high surface recombination velocities S [59, 60]. In this case we relate the effective charge carrier lifetime with the front and rear recombination employing the transcendental equation

$$tan(\alpha_0 W) = \frac{S_{front} + S_{rear}}{\alpha_0 D - \frac{S_{front} S_{rear}}{\alpha_0 D}} \qquad (2.5)$$

with

$$\alpha_0 D = \frac{1}{\tau_{eff}} - \frac{1}{\tau_{bulk}} \qquad (2.6)$$

to calculate S_{front} and S_{rear}. Here D denotes the minority charge carrier diffusion coefficient. This equation is valid in the case of transient lifetime measurements, where the time dependence of the excess charge carrier lifetime is evaluated. Note, that Eqn. 2.5 is also a good approximation in the case of static lifetime measurements [17].

Using the approximation in Eqn. 2.4 we relate now the surface recombination velocity to the recombination current J_{rec} by

$$S = q\frac{J_{rec}}{\Delta n} \qquad (2.7)$$

where q denotes the elementary charge and Δn the local excess charge carrier density at the interface. In low level injection

2.2. Dynamic Infrared Lifetime Mapping (ILM)

$$\Delta n \sim \frac{n_i^2}{N_A} \left[exp\left(V_i/V_{th}\right) - 1 \right] \qquad (2.8)$$

holds with the base doping concentration N_A and the intrinsic charge carrier concentration n_i. Thus we obtain from the Eqn. 2.1, 2.7 and 2.8 that the reverse saturation current density

$$J_{0,i} = q \frac{n_i^2}{N_A} S_i \qquad (2.9)$$

of a surface i is proportional to its surface recombination velocity S_i. This relationship allows us to determine recombination properties of a solar cell interface from lifetime measurements.

2.2. Dynamic Infrared Lifetime Mapping (ILM)

In the previous section we demonstrated the equality of recombination currents in a solar cell described by a reverse saturation current density J_0 and recombination currents in a lifetime measurement described by a surface recombination velocity S. We therefore employ in this work lifetime measurements to quantify contact recombination.

For this purpose we employ the dynamic Infrared Lifetime Mapping (ILM) technique [16]. This lifetime measurement method is calibration-free yielding images of the effective charge carrier lifetime τ_{eff} of silicon wafers within seconds. Since only the time dependence of the detected camera signal is used to determine τ_{eff} it is possible to measure also metallized samples. This makes dynamic ILM ideal for the measurement of the contact recombination velocity S_{cont}.

2.2.1. Setup

The sample under test is placed on an aluminum mirror as sketched in Fig. 2.1. The mirror heats the sample to a temperature of $70°C$. The lifetime measurement at elevated temperatures of $70°C$ allows in the present setup to increase the signal-to-noise ratio. Excess charge carriers are excited in the sample by illumination with LED arrays emitting photons at a wavelength of $930\ nm$. The infrared camera detects the change in free carrier emission in the sample due to a change in the excess charge carrier density.

Making use of the proportionality between the infrared light emission and the free charge carrier density inside the sample, an image of the effective charge carrier lifetime τ_{eff} is obtained by applying a lock-in technique.

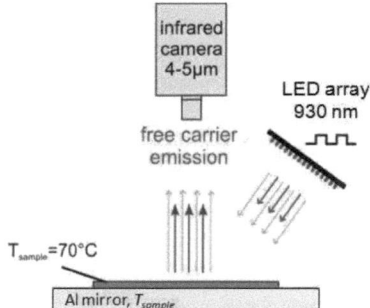

Figure 2.1.: *Sketch of the dynamic ILM setup taken from Ref. [61]. Charge carriers are excited in the sample by square-wave-shaped illumination with diode arrays. During the measurement, samples are placed on a temperature controlled infrared mirror. The infrared emission of the free charge carriers is recorded by an infrared camera.*

2.2.2. Principle

The excess charge carrier density Δn as a function of time t follows the continuity equation

$$\frac{d}{dt}\Delta n = G - \frac{\Delta n}{\tau} \tag{2.10}$$

where G is the generation rate of excess charge carriers due to illumination. Using a frame rate of 160 Hz and a lock-in frequency of 40 Hz, we take 4 pictures in one lock-in period of $T = 25\ ms$ duration. We apply a cosine and sine correlation in our lock-in approach. As a result we obtain two images $S_{cos} = S_1 - S_3$ and $S_{sin} = S_2 - S_4$, where S_i equals the signal integrated over the time t_{int} in image i. Measuring the ratio

$$\frac{S_{cos}}{S_{sin}} = \frac{t_{int} - 2\tau_{eff}[1 - exp(-\frac{t_{int}}{\tau_{eff}})]}{t_{int} - 2\tau_{eff}[exp(-\frac{T}{4\tau_{eff}}) - exp(-\frac{T+4t_{int}}{4\tau_{eff}})]} \tag{2.11}$$

we determine the effective charge carrier lifetime τ_{eff} using the known quantities t_{int} and T. Note, that in practice a more advanced evaluation technique is used. For this purpose we consider a non ideal square-wave shaped illumination following the arguments in Ref. [61].

2.2. Dynamic Infrared Lifetime Mapping (ILM)

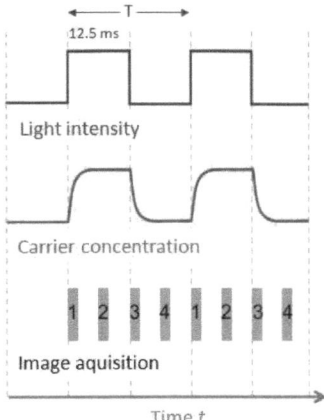

Figure 2.2.: *Schematic of the lock-in principle used for the dynamic ILM approach (taken from [62]). The time dependence of the incident light intensity and the charge carrier concentration are displayed together with the image acquisition times.*

2.2.3. Calibration of static ILM images

Even though the principle of the dynamicILM is rather simple, several effects preventing a precise evaluation of the effective lifetime τ_{eff} have been identified in Ref. [61]. The two most important effects are:

- an increase in the evaluated lifetime due to a sample temperature change especially in low lifetime regions and

- internal reflections of the emitted infrared light in the sample causing blurring.

When employing steady-state ILM [63] both effects are negligible. This is demonstrated in Fig. 2.3 by comparing the effective lifetime images of a locally point contacted sample obtained by dynamic and steady-state ILM. The steady-state ILM is calibrated in terms of excess charge carrier density by comparison with the dynamic ILM in a high lifetime region [61]. Note, that the uncalibrated steady-state ILM image does not require an additional measurement. It is however equal to $S_{sin} = S_2 - S_4$ which is already delivered in the lock-in approach as explained in the previous section.

We observe for the dynamically calibrated steady-state ILM a significantly improved sharpness of the image, compared to the dynamic ILM measurement in Fig. 2.3. The

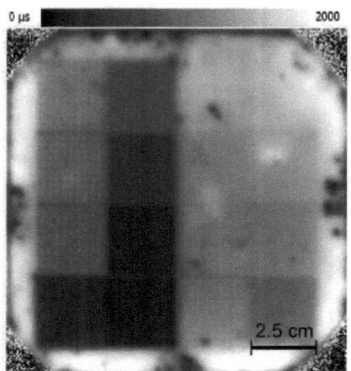 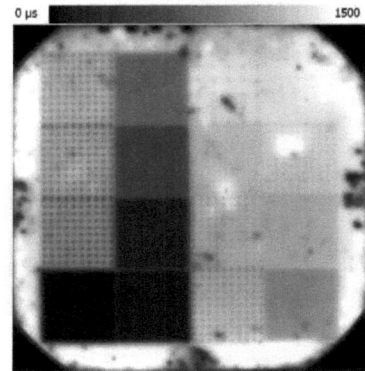

Figure 2.3.: *Comparison of dynamic ILM (left) and dynamically calibrated steady-state ILM (right). In the dynamically calibrated steady-state ILM image both, an increase in the evaluated lifetime due to a sample temperature change and internal reflections causing blurring can be neglected.*

impact of internal reflections is much less pronounced when using steady-state ILM which effectively reduces blurring. Furthermore we observe in the bottom left corner of the wafer a low lifetime region. Here the signal contribution due to a sample temperature change is quite distinct. As a result we evaluate with the dynamic ILM technique a significantly increased lifetime of $\tau_{eval} = 291\mu s$. However, using the dynamically calibrated steady-state ILM this effect is negligible allowing for the determination of the correct lifetime of $\tau_{eff} = 43\mu s$. Due to the significant improvements in lateral resolution as well as the precision of the evaluated lifetime we use throughout this work the combination of dynamic and steady-state ILM.

2.3. Separation of recombination in the passivated and contacted area

So far we developed an understanding of the recombination current to a surface of a solar cell. Furthermore we are able to measure the recombination current employing dynamic ILM measurements known from the previous section. As we consider now a sample with full area metal rear contact, we expect the recombination current to be homgenous and perpendicular to the metallized surface. This makes it easy to determine the contact recombination velocity S_{cont} from lifetime measurements.

2.3. Separation of recombination in the passivated and contacted area

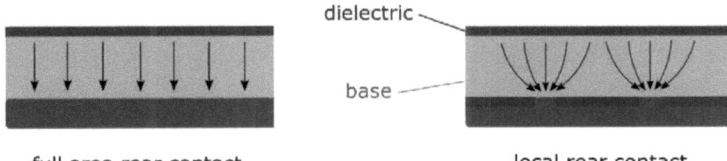

Figure 2.4.: *Sketch of the recombination current in a lifetime sample with full area (left) and local contacts (right). In the case of local contacts we expect strong lateral current flow.*

In the case of local contacts the situation considerably changes. The recombination velocity S_{cont} at the contacts is large compared to the surface recombination velocity in between. The homogenously created excess charge carriers will therefore mainly recombine at the interface to the local contacts. As a result we obtain in contrast to a fully contacted area an inhomogenous recombination current flowing perpendicular and lateral to the contacted surface as depicted in Fig. 2.4. This makes it difficult to discriminate between the recombination at the contacts and in between.

Models describing transport in a locally contacted sample are rather complex. However, we need an understanding of current transport in a locally contacted sample to interpret our lifetime measurements in terms of S_{cont} correctly.

Finite element simulations were often used to solve the transport equations and to optimize the local contact geometry [21, 22, 24]. However, they have the disadvantage that a simulation program is needed and furthermore each specific geometry of interest has to be simulated in a tedious procedure.

As an alternative to numerical simulations Rau et al. [23] developed an analytic solution for periodic contact geometries by solving the three dimensional transport equations in fourier space. Another approach from Cuevas is based on iterative calculations [64, 65]. However, in this work we employ an even more basic analytic approach which is often referred to as the "Fischer model" originally presented in the PhD thesis of Bernhard Fischer [17]. It allows to describe the three dimensional charge carrier transport in a locally contacted sample with a one dimensional equivalent.

2.3.1. Derivation of the Fischer model

The 'Fischer model' is a one dimensional analytic approximation of minority charge carrier transport in a locally contacted sample [17]. It allows us to distinguish between recombination at the local contacts and at the passivated interface in between the contacts. We will therefore use this model to determine the contact recombination velocity S_{cont} from measurements of an effective surface recombination velocity $S_{eff,rear}$. Furthermore we will use the Fischer model to gain a deeper understanding of recombination at local contacts which is possible to its basic analytic and one dimensional approach. This will allow us to identify possible restrictions of our method to determine S_{cont}.

In the following we only sketch the basic assumptions for the derivation of the Fischer model. The model is based on a decoupling of charge carrier transport equations in a solar cell. First we restrict ourselves to consider transport in the base only. According to Fischer [17] the disregarded emitter will act in three ways:

- Emitter recombination described by $J_{0,e}$ will add as a contribution to the total diode reverse saturation density J_0.

- The incomplete collection of carriers generated within the emitter will lead to a short circuit current loss.

- A lateral voltage drop due to the sheet resistance will act to provide a non-constant potential at the space charge region. This can be described by a contribution to the series resistance of the solar cell, so the front side may be treated as a constant potential surface.

All three emitter effects can be included in a solar cell model in a later stage.

For the description of transport in the base low level injection conditions are assumed, which allows to decouple minority and majority charge carrier transport. The majority charge carrier flow will experience a series resistance that is determined separately by solving the Laplace equation for the electrostatic potential. Fischer also neglected bulk recombination. Hence, only the minority charge carrier transport by diffusion is left to be described.

The diffusion equation is now solved by using the formal equivalence between diffusion equation and poisson equation to solve for the electrostatic potential. This allows to relate the effective diffusion length in the base to the series resistance in the base. Now analytic approximations for the series resistance of a local contact pattern comprising a) equally spaced point contacts taken from Cox and Strack [66] and b) parallel line contacts taken from Gelmont and Shur [67] allow to describe the base recombination analytically.

2.3.2. The effective rear surface recombination velocity

From the aforementioned assumptions Fischer derived an area averaged effective rear surface recombination velocity

$$S_{eff,rear} = \left(\frac{R_s - \rho W}{\rho D} + \frac{1}{fS_{cont}} \right) + \frac{S_{pass}}{1-f} \qquad (2.12)$$

2.3. Separation of recombination in the passivated and contacted area

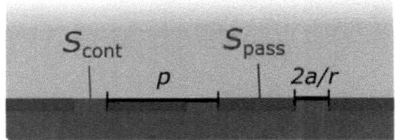

Figure 2.5.: *Recombination at local contacts of distance p is described by the contact recombination velocity S_{cont}. In between the contacts a passivating layer features a much lower recombination velocity S_{pass}. The contact size is described by a) the radius r in the case of point contacts and b) the line width a in the case of line contacts.*

as a function of the base resistivity ρ, the contact layout specific base series resistance R_s, the metallization area fraction f, the sample thickness W and the surface recombination velocity in the passivated area S_{pass} as well as under the contact S_{cont}. This relationship between S_{cont} and $S_{eff,rear}$ allows us to determine S_{cont} from lifetime measurements with the dynamic ILM technique. The sketch in Fig. 2.5 gives an overview over the model parameters.

The metallization fraction of point

$$f_{points} = \frac{\pi r^2}{p^2} \tag{2.13}$$

and line contacts

$$f_{lines} = \frac{a}{p} \tag{2.14}$$

is a function of the contact spacing p as well as the point radius r or the line width a. We calculate the base series resistance

$$R_{s,points} = \frac{p^2 \rho}{2\pi r} arctan\left(\frac{2W}{r}\right) + \rho W \left(1 - exp\left(-\frac{W}{p}\right)\right) \tag{2.15}$$

of a point contacted sample according to Cox and Strack [66]. For a line contacted sample we use the expressions introduced by Gelmont and Shur [67]:

$$R_{s,lines} = \frac{p\rho}{2\pi} ln\left(\frac{2\left(\sqrt{cosh\frac{\pi a}{4W}+1}\right)}{\sqrt{cosh\frac{\pi a}{4W}-1}}\right) + \rho W\left(1-exp\left(-\frac{W}{p}\right)\right) \tag{2.16}$$

when $tanh\frac{\pi a}{4W} \leq 1/\sqrt{2}$ and

$$R_{s,lines} = \frac{p\rho\pi}{2}\left[ln\left(\frac{2\left(1+\sqrt{tanh\frac{\pi a}{4W}}\right)}{1-\sqrt{tanh\frac{\pi a}{4W}}}\right)\right]^{-1} + \rho W\left(1 - exp\left(-\frac{W}{p}\right)\right) \quad (2.17)$$

when $1/\sqrt{2} < tanh\frac{\pi a}{4W} < 1$.

Note, that the equations for the base series resistance R_s are only approximations. For specific contact geometries such as very thin solar cells with a thickness below $50\mu m$ other equations for R_s are necessary [25]. Furthermore they are only valid in the case of an equipotential front surface. As a result they may not apply to describe the base series resistance of a solar cell [24] especially in the case of large contact distances.

Despite the Fischer model also other analytic approximations for the recombination at local contacts exist [68, 69]. They differ only slightly in their respective derivation and also give similar results for practical cases. Saint-Cast et al. reviewed the general validity of those models [69]. He found that the Fischer model is applicable for small metallization fractions $f \ll 1$ and for a contact recombination velocity $S_{cont} \gg S_{pass}$ that is much larger in comparison to the recombination velocity in the passivated areas. However, these two prerequisites are usually fulfilled for Si solar cells. As the Fischer model is an analytic approximation which was often discussed in the literature we performed a validity analysis ourselves. It is further elaborated in the appendix in chapter A.

2.3.3. Restrictions for the determination of S_{cont}

Now, we analyze the impact of the parameters S_{cont} and f on the effective rear surface recombination velocity $S_{eff,rear}$. This analysis allows us identify possible restrictions of determining S_{cont} by measuring $S_{eff,rear}$. For this purpose we make some realistic assumptions to simplify the Fischer model in Eqn. 2.12.

We assume small contacts compared to the thickness of the wafer $W \gg r$ and the contact pitch to be large $p \gg W$ compared to the sample thickness. Considering a rather small metallization fraction $f \ll 1$ we find

$$S_{eff,rear} = \frac{4DS_{cont}f}{\pi r S_{cont} + 4D} + S_{pass} \quad (2.18)$$

Two relevant cases are to be distinguished here: First, $\pi r S_{cont}$ is much smaller or second, much larger than $4D$. In the first case we have

$$S_{eff,rear} - S_{pass} \sim S_{cont}f. \quad (2.19)$$

Hence, the area averaged $S_{eff,rear}$ is only determined by the product of S_{cont} times f. It does not depend on the diffusion coefficient D since transport is not a limiting factor in this case. For a determination of S_{cont} with Fischer's model it is therefore essential to determine the metallization fraction f independently.

In the second case we have

$$S_{eff,rear} - S_{pass} \sim \frac{4Df}{\pi r}. \tag{2.20}$$

The recombination at the sample rear is then independent of S_{cont}. In this case recombination is transport limited and thus only dependent on geometry parameters such as p and, of course, the diffusion coefficient D. Note, that in this case an accurate determination of S_{cont} with Eqn. 2.12 is not possible.

2.4. Application example

In this section we give an application example on how to extract the contact recombination velocity S_{cont} from lifetime measurements. We published this work in Ref. [58]. We apply the analytic 'Fischer model' presented in section 2.3.1 to lifetime measurements obtained from dynamically calibrated steady-state ILM. For this purpose we prepare samples with two different local contact types:

- **Local Contact Opening (LCO)** - Contact is made by laser ablation of the dielectric layer and subsequent evaporation of Al [18].

- **Laser Fired Contacts (LFC)** - First Al is evaporated and then Si is contacted through the dielectric layer by laser [19].

We apply our technique for the determination of S_{cont} first to LCO and LFC instead of local Al alloyed contacts. This has the following advantage: As introduced in chapter 1 local Al alloyed contacts have a three dimensional structure forming structures deep in the Silicon bulk. Furthermore they are composed of different layers, that is the eutectic and the Al-p^+ layer. Such structures are not explicitly considered in the Fischer model. For this purpose we test our technique at more simple to describe local contact types: the LCO and LFC.

2.4.1. Sample preparation

We study the contact recombination as a function of the base doping concentration. For this purpose we prepare (100) oriented floatzone (FZ) silicon samples with specific resistances of 0.6, 1.4, 3.9 and 230Ωcm. Etching in KOH reduces the initial thickness of the wafers to approximately 250μm, before applying a standard RCA clean followed by a HF dip.

Both surfaces are electronically passivated by a stack of 10nm amorphous silicon (a-Si) and 100nm silicon nitride (SiNx). The latter has a refractive index of $n = 1.9$. We deposit

Table 2.1.: *Pitch p and metallization fraction f_{LCO} of the LCO geometries, which are identified in Fig. 2.6. The same pitches p are used for the LFC geometry.*

No.	Pitch p [mm]	Metallization fraction of LCOs f_{LCO}
1	2.5	4.4×10^{-3}
2	2.5	6.4×10^{-3}
3	2.5	1.1×10^{-2}
4	0.05	5.5×10^{-1}
5	0.75	2.4×10^{-3}
6	0.5	5.3×10^{-3}
7	0.28	1.8×10^{-2}
8	0.2	3.6×10^{-2}
9	5.0	5.5×10^{-5}
10	2.0	3.5×10^{-4}
11	2.5	6.8×10^{-4}
12	2.5	1.7×10^{-3}
13	-	no contacts
14	2.5	2.2×10^{-4}
15	1.5	6.2×10^{-4}
16	1.0	1.4×10^{-3}

both layers by plasma enhanced chemical vapour deposition, using a parallel plate reactor for a-Si deposition at $235°C$ and an inline remote reactor for SiN deposition at $400°C$.

We apply two different contacting schemes:

- Laser contact openings (LCOs) formed by ablation of the dielectric stack and subsequent deposition of $2\mu m$ Al, using an inline physical vapour deposition (PVD) tool. As has been shown in Ref. [70], the surface recombination properties of the contact are not influenced by the parameters of the metallization process.

- Deposition of $10\mu m$ Al in the same inline PVD deposition tool, followed by firing the contacts through the passivation layers (LFCs). The LFC technique applied at ISFH requires the deposition of $10\mu m$ Al in order to achieve an appropriate contact formation.

While a 532 nm laser system with a pulse length of approximately 10 ps (Gaussian profile) is used for the LCOs, the LFCs are formed with laser pulses at 532 nm that have an approximate length of 20 ns (Gaussian profile).

2.4. Application example

In order to investigate the local contact recombination, we use 16 different contact geometries. The corresponding contact pitches p and metallization fractions f of LCOs are listed in Table 2.1. The variation in size of the contacts is achieved by overlapping single point contacts. A matrix of overlapping single point contacts shapes a large contact. The metallization area fraction

$$f = \frac{ab}{p^2} \qquad (2.21)$$

of these rectangular-shaped contacts depends on the edge lengths a and b and the contact pitch p. Large contacts have been applied with constant pitch $p = 2.5mm$ and differing contact size $a \times b$.

2.4.2. Effective rear surface recombination

We determine the effective rear surface recombination velocity $S_{eff,rear}$ of the 16 contact geometries to evaluate the contact recombination velocity S_{cont}. For this purpose we employ lifetime measurements using the dynamically calibrated steady-state ILM.

We perform the lifetime measurements at an illuminating photon flux of $3 \times 10^{16} cm^{-2} s^{-1}$ ensuring low level injection in all measurements. The a-Si/SiNx stack exhibits negligible injection dependence [22] and thus allows, evaluating all 16 squares with just one measurement. The use of high quality defect-free FZ silicon justifies, to assume a limitation of the bulk lifetime τ_{bulk} by radiative and Auger recombination. We use the parametrization of Kerr [71] to calculate the bulk lifetime. All dynamic ILM measurements have been performed at $70°C$ sample temperature, to obtain a high signal to noise ratio at short measurement times. The difference of the evaluated lifetime is below 10 %, when comparing lifetime measurements at $35°C$ and $70°C$.

We observe in our lifeteime images a homogeneous lifetime distribution in squares #4 to #8 and #16 as exemplarily demonstrated in Fig. 2.6. The numbers in Fig. 2.6 identify the geometries, which are applied to an area of $2.5 \times 2.5 cm^2$. The homogenous lifetime distribution is the result of the small contact pitch p compared to the wafer thickness W [72]. However, some squares in Fig. 2.6 feature an inhomogeneous lifetime distribution with low lifetimes at the contact and high lifetimes in between.

In the case of rather low surface recombination velocities S_{front} and S_{rear} the approximation

$$\frac{1}{\tau_{eff}} = \frac{1}{\tau_{bulk}} + \frac{S_{front} + S_{rear}}{W}. \qquad (2.22)$$

holds [59]. We therefore use the harmonic mean $\langle \tau_{eff}^{-1} \rangle^{-1}$ to average the effective charge carrier lifetime. Now S_{pass} follows from Eqn. 2.5, assuming $S_{front} = S_{rear}$ in square #13

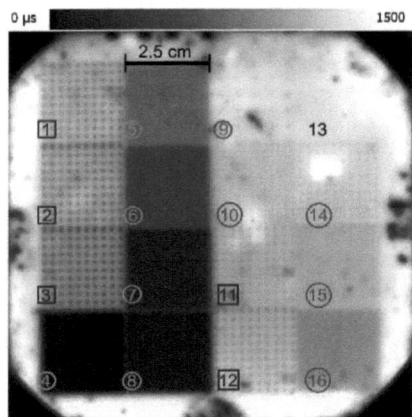

Figure 2.6.: *Dynamically calibrated steady-state ILM of one of the wafers used for the analysis. Applied are LCOs in 16 different contact geometries. Numbers in circles denote areas with single point contacts and numbers in squares identify areas with large contacts. No contacts are applied to area no. 13.*

without contacts. Subsequently, we use Eqn. 2.5 to calculate $S_{eff,rear}$ of the 15 squares with metal contacts.

2.4.3. Contact recombination velocity

We use Eqn. 2.12 to determine the contact recombination velocity S_{cont} from the measured effective rear surface recombination velocity $S_{eff,rear}$ and the recombination velocity in the passivated area S_{pass}. However, in Eqn. 2.12 we find also parameters depending on the local contact geometry.

As a result the contacted area fraction f has to be determined independently of the lifetime measurements, to determine S_{cont} accurately. Using an optical microscope, we determine the contact pitch p and the contact size. Figure 2.7 shows optical microscope images of a single point LCO (left) and a large LCO (right), prior to metal deposition. The area of ablated dielectric layer appearing yellow in Figure 2.7 is used, to calculate the radius of a circle with equal area in the case of single point contacts. A contact radius of $r_{LCO} = 21\mu m$ is measured for LCOs. In the case of large contacts the edge lengths a and b are measured. The area of ablated dielectric layer is elliptical, due to the elliptical Gaussian laser beam profile. Hence the overlap of ablated area at large contacts is better in the horizontal direction, than in the vertical direction in Fig. 2.7.

2.4. Application example

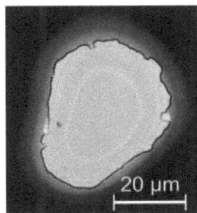 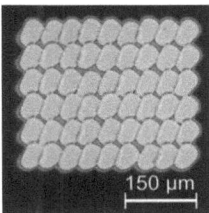

Figure 2.7.: *Micrograph of a single point LCO (left) and a large LCO (right) before metallization. The dieletric stack of $a-Si$ and SiN_x appears dark blue, while the Si surface is bright yellow.*

We investigate the effective rear surface recombination velocity $S_{eff,rear}$ as a function of the metallization fraction f of single point and large LCOs at $N_A = 7 \times 10^{13} cm^{-3}$ in Figure 2.8. Single point LCOs are applied with constant radius r and large LCOs with constant pitch p. From the data in Fig. 2.8 we determine the contact recombination velocity S_{cont} using the analytical Fischer model in Eqn. 2.12.

Employing a least squares fit to the data we determine $S_{cont} = 10^4 cm/s$ for single point LCOs. In contrast, the same analysis made for the large contact data results in $S_{cont} = 10^3 cm/s$. Plotting the analytical model with $S_{cont} = 10^4 cm/s$ and constant pitch $p = 2.5mm$ in Fig. 2.8 (blue line), confirms the difference in S_{cont} of single point and large LCOs. However, a decrease in S_{cont} from single point contacts to large contacts is unlikely, since the area of remaining passivation within the large contact is negligible and recombination properties are not expected to change by overlapping of single point contacts.

The microscopic images of the LCOs in Fig. 2.7 show a distinct contrast in the vicinity of the contact, compared to the surrounding area. This contrast suggests a modification of the passivation stack and hence of the passivation quality around the ablated area. As the used laser features an elliptical Gaussian beam profile, the modification outside of the ablated area is possibly introduced by laser light. Another possible explanation is mechanical stress in the dielectric layer, which is introduced by the ablation process itself [73, 74]. However, scanning electron microscopy images (not shown) indicate no change in structure of the passivation stack.

When comparing the two images in Fig. 2.7, it becomes clear that a peripheral defected area would have a smaller impact on the recombination of large contacts, compared to single point contacts. Thus, with increasing size of the large contacts, the ratio of the defected area to the large contact area decreases. As a result we expect the impact of the laser-damaged area to become negligible, when analysing large contacts of sufficient size.

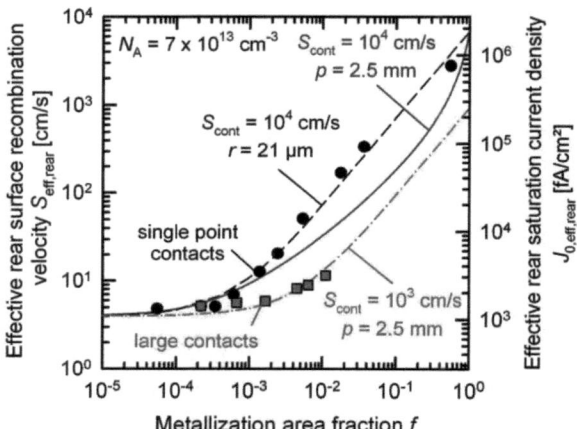

Figure 2.8.: *Measured effective rear surface recombination velocity $S_{eff,rear}$ at one of the samples with single point (black circles) and large LCOs (red squares) as a function of the metallization area fraction f. The lines show a fit of the analytic Fischer model with Eqn. 2.12 to the data, with constant radius r (black dashed line) and constant pitch p (red dashed dotted line). Note, $S_{eff,rear}$ depends on both f and p (blue solid line).*

A laser-damaged peripheral area at single point contacts would result in an overestimation of S_{cont}, since the recombination occurring in the laser damaged area would be assigned to the contact area. Level and distribution of laser damage in the vicinity of the contacts is unknown. Thus, we assume equal surface recombination velocities in the contacted and degraded area for simplicity. Estimating an effective single LCO radius $r_{eff,LCO} = 27\mu m$ from the left microscopic image in Fig. 2.7, yields $S_{cont} = 8 \times 10^3 cm/s$. One might also assume $S_{cont} = 10^3 cm/s$ as for large contacts. This would result in an effective single LCO radius of $r_{eff,LCO} = 55\mu m$.

To investigate this topic further, a measurement of the excess carrier density around one point contact with μm resolution and a model of the local charge carrier transport are required. However, this is beyond the scope of our investigation. Since no direct experimental proof of a laser-damaged peripheral area has been found, we present S_{cont} values of single point contacts, without considering any laser damage and thus assuming homogeneous recombination properties within the circular contact. This is an assumption that may not be fulfilled in reality.

2.4.4. Impact of the doping density

In the following S_{cont} of single point contacts and large contacts is determined for samples with different base doping concentrations and for both contact formation schemes, LCOs and LFCs. The corresponding reverse saturation current densities J_0 are obtained using Eqn. 2.9. To allow for an estimate of the accuracy of the investigation, a measurement error of 20 % has been assumed for all surface recombination velocities, which accounts for 10 % systematic error and 10 % statistical error.

The recombination properties of the contact are determined, by a least squares fit of the Fischer model in Eqn. 2.12. An upper limit for the surface recombination S_{cont} constitutes the thermal velocity of minority charge carriers in Si $v_{th} \sim 10^7 cm/s$ [26], limiting the supply of recombining charge carriers, even in the case of an infinite surface recombination velocity. The results of this analysis are shown in Fig. 2.9 together with literature data.

In the left part of Fig. 2.9 the dependence of S_{cont} on the doping density N_A is shown for single point contacts and large contacts, consisting of LCOs and LFCs. The measured S_{cont} at single point contacts and large contacts of our LFCs show no significant difference to LCOs. A difference between single point contact and large contact S_{cont} values of at least one order of magnitude can be observed for every investigated N_A, similar as in Fig. 2.8. While we cannot measure an impact of N_A on S_{cont} at large contacts within the measurement accuracy, S_{cont} at single point contacts varies over nearly three orders of magnitude depending on N_A. However, the error bars suggest, that in the case of these high surface recombination velocities, an accurate determination of S_{cont} is impossible. In section 2.3.1 we found that in this case the recombination at the contact is not anymore

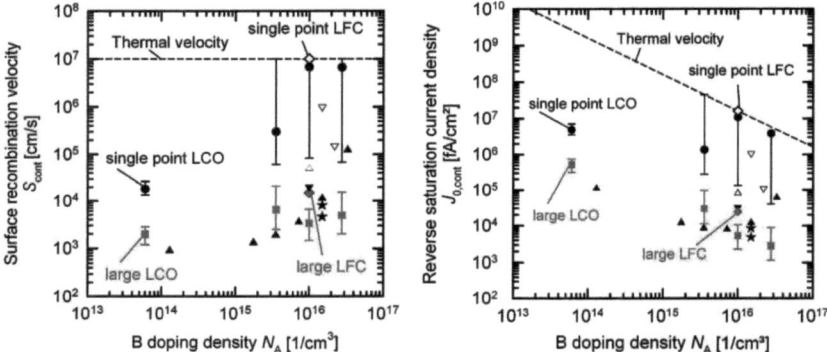

Figure 2.9.: *Measured contact recombination velocity S_{cont} (left) and reverse saturation current density $J_{0,cont}$ of different contact formation and measurement techniques. The data obtained in this study at LCOs and LFCs results from the analysis of single point contacts (black circles, open rhombus) and large contacts (red squares, filled rhombus). These values are compared with literature data (open triangle down Schoefthaler et al. [72, 75] - photolithographically opened point contacts; filled triangle down Schmidt [27] - line contacts opened by metal evaporation prior to passivation; open triangle up Plagwitz and Brendel [68] - COSIMA; filled triangle up Kray and Glunz [76] LFC; Stars Nekarda et al. [77] - LFC). The thermal velocity v_{th} of charge carriers in Si (black dashed line) is the upper limit for the contact recombination.*

influenced by S_{cont} since $S_{cont} \gg 4D/(\pi r) \sim 10^4 cm/s$.

The S_{cont} values reported by other studies are in between the S_{cont} values, which we find for point contacts and large contacts. Kray et al. [76] and Nekarda et al. [77] investigated LFCs, while Schoefthaler et al. [72,75] applied photolithographically opened point contacts. Schmidt [27] analyzed line contacts by metal evaporation prior to passivation and Plagwitz et al. [68] used COSIMA point contacts.

According to the data provided by these studies, LFC contacts exhibit a lower contact recombination velocity S_{cont} compared to the other contact formation techniques. This has been attributed to the formation of a highly doped Al region induced by the LFC process [35, 76].

The existence of such a highly doped region has been impressively proven by using the LFC technology for emitter formation [34] and is hence expected to lower contact recombination according to cahpter 1. However, our results show no difference in terms of surface recombination velocity between LFCs and LCOs. We expect that the equality of contact recombination at LCO and LFC may be the result of a deep crystal damage during ns laser processing [78], which compensates for the effect of the highly doped layer.

In the right part of Fig. 2.9 the same data is plotted after converting the surface recombination velocities into reverse saturation current densities using Eqn. 2.9. The most striking result, is the linear dependence of $J_{0,cont}$ on N_A, for LCOs and LFCs obtained for large contacts in this study. While reverse saturation current densities at diffused surfaces, like pn-junctions, are commonly assumed to be independent of the doping density, this is obviously not the case for the data in Fig. 2.9.

2.5. Conclusion

In this chapter we introduced how the recombination current J_0 of an interface in a solar cell is assessed experimentally by measuring the surface recombination velocity S using charge carrier lifetime measurements. For this purpose we employ the dynamic Infrared Lifetime Mapping (ILM) technique which allows to evaluate the effective lifetime of metallized samples analyzing only the time dependence of the measured signal. To enhance spatial resolution and sensitivity at low lifetimes we use static ILM images calibrated using the dynamic ILM technique. Analyzing those images employing an analytical model of the three dimensional charge carrier transport in a locally contacted sample, we are able to separate recombination in the passivated and contacted areas. From a deeper analysis of the used model we deduce that a precise determination of the contact recombination is not possible when charge carrier transport is transport limited.

We test the proposed technique for the measurement of the contact recombination velocity S_{cont} at LCO contacts and measure local reverse saturation current densities as low as

$J_{0,cont} = 2 \times 10^3 \, fA/cm^2$ at those contacts. We observe no difference in $J_{0,cont}$ between LCO and LFC which is in contrast to the general believe that recombination at LFCs is reduced due to a highly doped layer underneath the contacts. Our results indicate degradation of the passivation stack due to laser treatment in the vicinity of the LCO and LFC.

3. Recombination at local Al alloyed contacts

While the basic concept of local Al alloyed contacts was already introduced in chapter 1, we now analyze the recombination properties of local Al alloyed contacts. For this purpose we employ the technique presented in the previous chapter 2 to measure the contact recombination velocity S_{cont}. We prepare samples with varying contact geometry and measure the effective rear surface recombination velocity $S_{eff,rear}$. Using the analytical model presented in chapter 2 we determine the contact recombination velocity S_{cont}. We prove the assumption of reduced contact recombination due to the formation of a highly Al doped Al-p^+ layer beneath the local contacts, using scanning electron microscopy (SEM) images with secondary electron contrast. We confirm this understanding with PC1D simulations. This work is published in Ref. [45].

3.1. Sample preparation

For our analysis we prepare lifetime samples in the following way:

We use (100)-oriented p-type float-zone (FZ) silicon wafers of 1.5 and 200 Ωcm resistivity, with a thickness of 275 μm after saw damage etching in KOH. The samples are pseudosquare with a size of $12.5 \times 12.5 cm^2$. We use 200 Ωcm FZ material as it features high bulk lifetimes in the ms range allowing to measure the lifetime accurately and without effects from insufficient surface passivation. The 1.5 Ωcm is however more comparable to silicon used for solar cell production and is therefore also used in this study.

Subsequent to a RCA clean followed by a HF dip, both nearly planar wafer surfaces were electronically passivated by a stack of 17 nm thermal oxide (SiO_2) and 75nm silicon nitride (SiN_x). The oxide was grown in dry O_2 ambient at 900°C. Using a parallel plate reactor, we deposited the SiN_x layers by plasma enhanced chemical vapour deposition at 400°C on both surfaces.

We apply single sided local contact openings (LCO) by laser ablation of the dielectric stack. As shown in Fig. 3.1 we use two different contacting schemes: a) parallel line contacts and b) equally spaced point contacts. We realized the openings of radius r and line contacts of line width a by overlapping single openings of 25μm radius. In order to

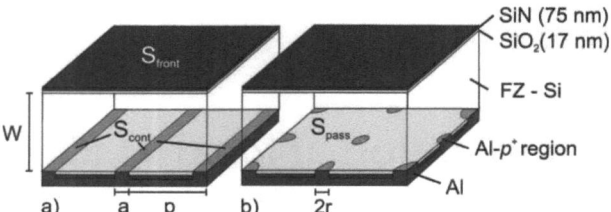

Figure 3.1.: *Schematics of the sample structure under test (not to scale). A dielectric stack consisting of 17 nm SiO_2 and 75 nm SiN on top is applied on both sides of FZ-Si wafers of 1.5 and 200 Ωcm resistivity. Laser contact openings (LCOs) are realized by laser ablation of the rear dielectric stack in a) line and b) point geometries. The contact formation is subsequently realized, by full-area screen printing and firing of a standard fritless Al paste.*

investigate the local contact recombination the laser defined 16 contact geometries with an area of $2.5 \times 2.5\ cm^2$ on each $12.5 \times 12.5 cm^2$ pseudosquare sample. These vary in contact pitch p or contact size (i.e., radius r or line width a).

We complete the contact formation by full area screen-printing of a fritless Al paste on top of the LCOs. Subsequently, the samples were fired in an industrial conveyor belt furnace at approximately $800°C$ and conditions applicable to the cofiring of solar cells.

3.2. Contact recombination

To determine the local contact recombination velocity S_{cont} we follow the procedure in section 2.4.

3.2.1. Effective lifetime measurements

From effective charge carrier lifetime measurements employing dynamically calibrated steady-state ILM we determine the effective rear surface recombination velocity $S_{eff,rear}$ at each of the 16 local contact geometries. For this purpose we average the effective charge carrier lifetime τ_{eff} by using the harmonic mean $\langle \tau_{eff}^{-1} \rangle^{-1}$. Now S_{pass} follows from Eqn. 2.5, assuming $S_{front} = S_{rear}$ at square # 13 without contacts. Subsequently, we use Eqn. 2.5 to calculate $S_{eff,rear}$ of the 15 squares with metal contacts.

Symmetrically processed and fired reference samples without LCOs and without metallization are used, to analyze a potential degradation of the front passivation stack due to firing. We observe no lifetime degradation of the passivation stack. Additionally, the SiO_2/SiN_x stack exhibits negligible injection dependence [79]. On the 1.5 Ωcm material

3.2. Contact recombination

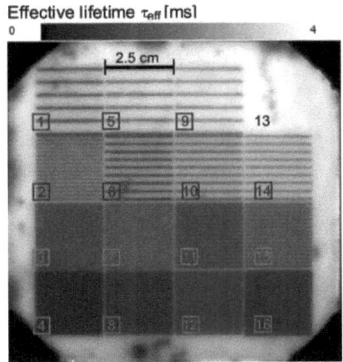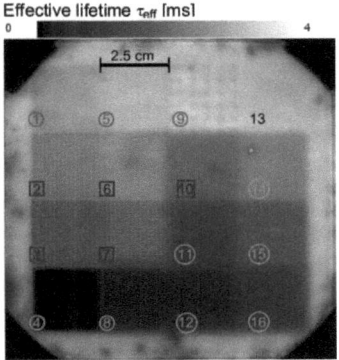

Figure 3.2.: *Dynamically calibrated steady state ILM images of samples with line (left) and point contacts (right), processed on FZ silicon with a resistivity of 200 Ωcm using the ps Laser. The numbers in circles denote areas with single point contacts and numbers in squares identify areas with large point or line contacts. No contacts are applied to area # 13.*

we measure in the injection range between $10^{14}cm^{-3} < \Delta n < 10^{15}cm^{-3}$ a lifetime variation smaller than 30%. Thus we are able to evaluate all 16 squares of different contact geometry with just one measurement.

Before we use the Fischer model in Eqn. 2.12 to calculate the contact recombination velocity S_{cont} we also need to measure the contacted area of the 15 applied local contact geometries. We determine the contact pitches p, contact radii r and line widths a employing micrographs of the LCOs before screen printing. They are comprised in Table 3.1. Note, that the contact size may increase during firing [80]. Hence, when using the contact sizes evaluated before contact firing the determined contact recombination velocities have to be regarded as upper limits of the actual ones.

3.2.2. Effective rear surface recombination velocity

The $S_{eff,rear}$ values determined from the lifetime measurements on $200\Omega cm$ substrates are shown in Figure 3.3 as a function of the metallization fraction f. The metallization fraction f is calculated from the original size of the LCOs measured with the light microscope. Employing the Fischer model in Eqn. 2.12 we obtain the contact recombination velocity S_{cont} by applying a least square fit to the data. Please, note that the contact size increases during firing (ref). In consequence all S_{cont} have to be considered as upper limits of the

Table 3.1.: *Pitch p and size (width/radius) of the point and line contact geometries, which are identified in Fig. 3.2*

No.	Line pitch $p_{Line}[mm]$	Line width $a[\mu m]$	Point pitch $p_{Point}[mm]$	Point radius $r[\mu m]$
1	5.0	250	2.0	25
2	1.5	80	1.5	150
3	1.0	135	1.5	240
4	0.6	250	0.05	25
5	5.0	135	2.5	25
6	3.0	250	1.5	100
7	1.0	80	1.5	195
8	0.6	135	0.23	25
9	5.0	80	5.0	25
10	3.0	135	1.5	50
11	1.5	250	0.75	25
12	0.6	80	0.23	25
13		no contacts		
14	3.0	80	1.5	25
15	1.5	135	1.0	25
16	1.0	250	0.5	25

3.2. Contact recombination

actual values when micrographs are used to determine the size of the contacts.

We determine contact recombination velocities as low as $S_{cont} = 90 cm/s$ at the line contacts when fitting the line contact data in Fig. 3.3 nearly independent of the line width. In comparison, the evaluation reveals $S_{cont} = 6 \times 10^4 cm/s$ for the single point contacts in Fig. 3.3. However, the analysis of the large point contacts with different contact sizes in Fig. 3.3 does not result in one unique S_{cont}. Here the large point contacts have constant pitch $p = 1.5 mm$ but differ in contact size. We calculate the radius r of noncircular-shaped point contacts using their contact area $A = \pi r^2$. At the largest point contact with $r = 240\ \mu m$, we measure $S_{cont} = 2 \times 10^2 cm/s$, which is two orders of magnitude smaller, as compared with the contact recombination velocity of the single point contacts with a radius of only $25 \mu m$. Hence, the data in Fig. 3.3 suggests a change in the recombination properties with contact size and geometry.

Note, that the measured recombination velocity at single point Al alloyed contacts $S_{cont,Alalloyed} = 6 \times 10^4 cm/s$ is of the same order of magnitude as the recombination velocity at the single point LCOs with $S_{cont,LCO} = 10^4 cm/s$ presented in chapter 2.

3.2.3. Contact recombination velocity

In order to verify that the recombination properties depend on the contact size and geometry, we investigate S_{cont} as a function of line contact width a and point contact radius r. For this purpose we use the same approach as used in Fig. 3.3 and apply it to the sample with $1.5 \Omega cm$ and $200 \Omega cm$ resistivity. The evaluated contact recombination velocities S_{cont} are shown in Fig. 3.4 as a function of the contact size, i.e. point radius and line width.

In our analysis we assume a measurement error of 20% for all $S_{eff,rear}$, which accounts for 10% systematic error and 10% statistical error. An upper limit for the contact recombination velocity S_{cont} constitutes the thermal velocity of minority charge carriers in Si $v_{th} \sim 10^7 cm/s$ [26], limiting the supply of recombining charge carriers, even in the case of an infinite surface recombination velocity.

When considering the point contact data, we observe a significant decrease in S_{cont} with increasing contact size in Fig. 3.4, independent of the material resistivity. Note, that the error in the determination of S_{cont} indicated by the error bars in Fig. 3.4 decreases with decreasing S_{cont}. In contrast to the point contact data, the S_{cont} values of the line contacts are found to be smaller. Furthermore they are nearly independent of the line width in the range $80 < a < 300 \mu m$.

Comparing the contact recombination velocity of the lower resistivity $1.5 \Omega cm$ material in the right diagram of Fig. 3.4 with those values obtained on the $200 \Omega cm$ material in the left diagram of Fig. 3.4, a significantly lower S_{cont} (around one order of magnitude) is found for the $200 \Omega cm$ material. This impact of the material resistivity on S_{cont} is comparable to the data obtained at LCO and LFC in section 2.4. The lowest contact recombination

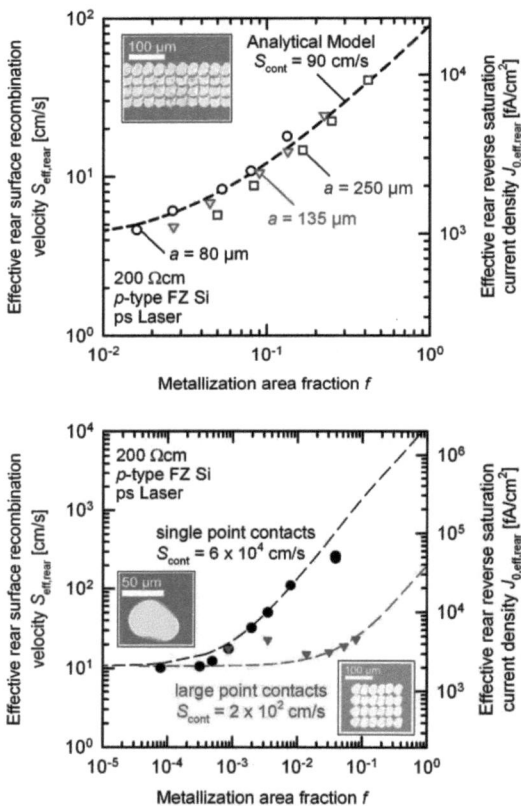

Figure 3.3.: *Effective rear surface recombination velocity $S_{eff,rear}$ and effective rear reverse saturation current density $J_{0,eff,rear}$ as a function of the metallization fraction f, measured at the samples in 3.2: a) line contacts of different line width and b) single point contacts of different pitch p (black circles) and large point contacts of different radius r (red triangles). The lines show the fit of the analytical model to the measured data. Line contacts and large point contacts are realized by overlapping of single point contacts, which is depicted in the small micrographs.*

3.2. Contact recombination

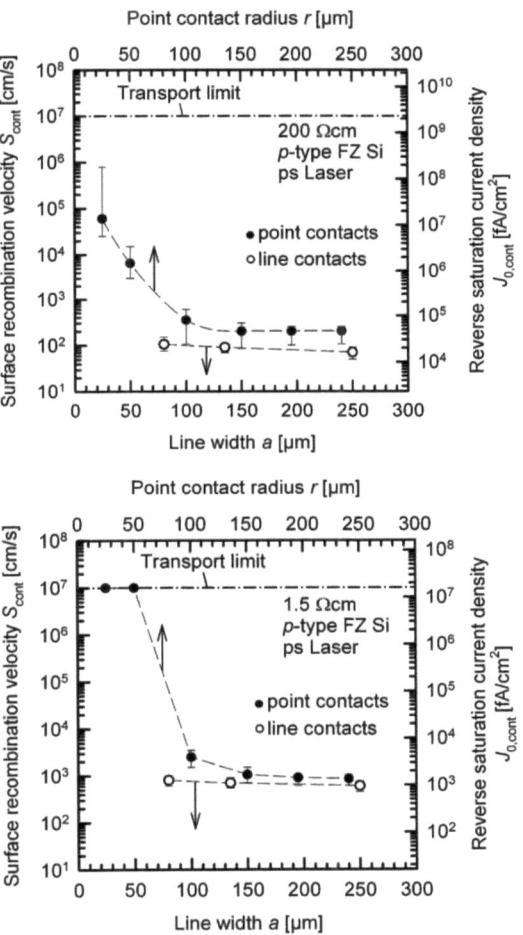

Figure 3.4.: *Contact recombination velocity S_{cont} and contact reverse saturation current density $J_{0,cont}$ at point contacts (closed circles) as a function of the contact radius and line contacts (open circles) as a function of the line width, measured on wafers of a) 200 Ωcm and b) 1.5 Ωcm resistivity. The lines are guides to the eye. The thermal velocity equals the transport limit.*

velocity of $S_{cont} = 65 cm/s$ (corresponding to $J_{0,cont} = 2 \times 10^4 fA/cm2$) is observed in Fig. 3.4 for line contacts of width $a = 250 \mu m$.

In contrast, the contact reverse saturation current density $J_{0,cont}$ is lower on the $1.5\Omega cm$ resistivity material. Values of $J_{0,cont}$ as low as $900 fA/cm2$ (corresponding to $S_{cont} = 600 cm/s$) are measured at the line contacts in Fig. 3.4. This value is comparable with those determined for a full-area Al alloyed contact [81]. As discussed in chapter 2 reverse saturation current densities are directly related to the recombination current in a solar cell. Hence, we can conclude that the recombination current at local Al alloyed contacts is smaller in the case $1.5\Omega cm$ material compared to the material of $200\Omega cm$ resistivity.

3.3. Structural investigation

In the previous section we measured an exciting impact of the contact geometry on the contact recombination. In this section we investigate the origin of this behaviour. In the case of full area Al alloyed contacts it was shown that low contact recombination is the result of a sufficiently thick Al-p^+ layer [33, 82] as described in chapter 1. A potential barrier induced by the highly doped layer prevents minority charge carriers to recombine at the metal interface.

It is assumed that the Al-p^+ layer observed at local Al alloyed contacts also lowers contact recombination. To prove this hypothesis we measure the contact recombination velocity S_{cont} as a function of the Al-p^+ layer thickness W_{Al-p^+}. The thickness of the local Al-p^+ layer is measured using scanning electron microscopy (SEM) images with secondary electron contrast.

3.3.1. Al-p^+ layer thickness

Fig. 3.5 shows a SEM image of a line contact together with a micrograph of the same contact, as captured before screen printing. The Al-p^+ region appears brighter in the SEM image than the high-resistivity bulk of the silicon wafer due to differing local ionization energy values [83]. The mean Al-p^+ layer thickness

$$W_{Al-p^+} = Area_{Al-p^+}/width_{Al-p+} \tag{3.1}$$

is determined by measuring the cross-sectional area $Area_{Al-p^+}$ of the Al-p^+ region and dividing it by $width_{Al-p+}$. Here $width_{Al-p+}$ denotes the approximate width of the Al-p^+ layer considering the contact to be the interface of Al-p^+ layer and Si, which is depicted in 3.5. The microscopic image allows identifying the line contact opening width a (as used for the evaluation of S_{cont}) to be $135 \mu m$. In contrast we find $width_{Al-p+} = 180 \mu m$ supporting the argument that the contact size of local Al alloyed contacts increases during firing [80].

3.3. Structural investigation

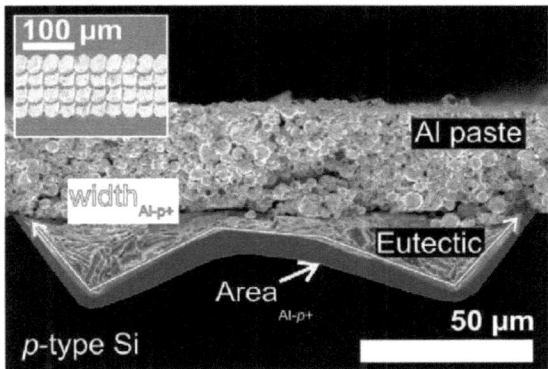

Figure 3.5.: *SEM image of a line contact. The Al-p^+ region appears brighter than the high resistivity bulk. In the upper left a micrograph of the contact is depicted.*

Using the above described procedure we evaluate the thickness of the Al-p^+ layer W_{Al-p^+} as a function of the contact sizes in Fig. 3.6. The most striking finding in Fig. 3.6 is a significant difference in the local Al-p^+ layer thickness W_{Al-p^+} for point and line contacts. Furthermore we find with increasing point contact radius an outstanding increase in W_{Al-p^+} from $W_{Al-p^+} = 0\mu m$ at $r = 50\mu m$ to $W_{Al-p^+} = 6.3\mu m$ at $r = 290\mu m$. Here the contact size is determined after contact formation. In the case of line contacts we observe an increase in W_{Al-p^+} from a line width of $a = 100\mu m$ to $a = 150\mu m$ after firing. However, increasing the line width further did not result in an increase of W_{Al-p+}. A detailed explanation of the observed W_{Al-p+} as a function of the contact geometry is given in chapter 4.

As the thickness of the Al-p^+ layer is considered to determine the contact recombination velocity [33], we now measure the contact recombination velocity S_{cont} as a function of the Al-p^+ layer thickness on $200\Omega cm$ FZ material in Fig. 3.7.

The data clearly indicates a decrease of S_{cont} with an increasing Al-p^+ layer thickness W_{Al-p^+}. Here, the contact recombination velocity S_{cont} decreases from $6 \times 10^4 cm/s$ at a W_{Al-p^+} of 0 to around $2 \times 10^2 cm/s$ at a W_{Al-p^+} larger than $2\mu m$. This behaviour is similar for point and line contacts. Hence we can state that for lowest contact recombination a mean W_{Al-p^+} of $2\mu m$ is necessary.

3.3.2. Predicting the contact recombination velocity

The relationship of contact recombination velocity S_{cont} and Al-p^+ layer thickness W_{Al-p^+} in Fig. 3.7 already demonstrates reduced contact recombination also at local Al alloyed

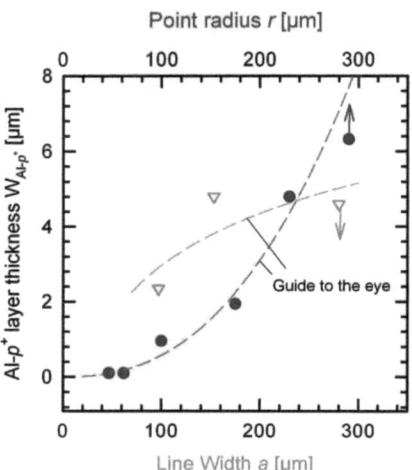

Figure 3.6.: *Al-p$^+$ layer thickness as a function of the contact geometry, i.e. the line width a (open orange triangles down) and point contact radius r (full violet circles).*

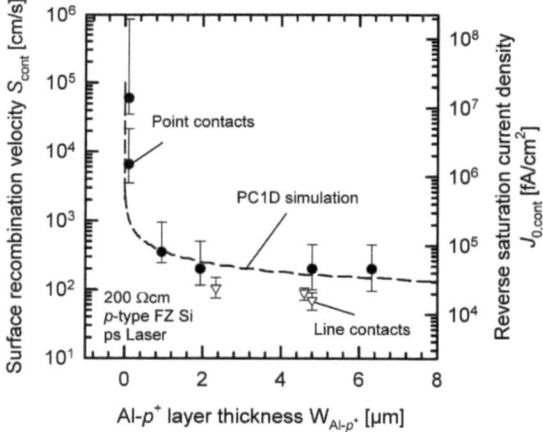

Figure 3.7.: *Surface recombination velocity S_{cont} as a function of the mean Al-p$^+$ layer thickness W_{Al-p^+} at ps Laser point (full circles) and line contacts (open triangle down). The PC1D simulation (dashed line) verifies the measured data.*

3.3. Structural investigation

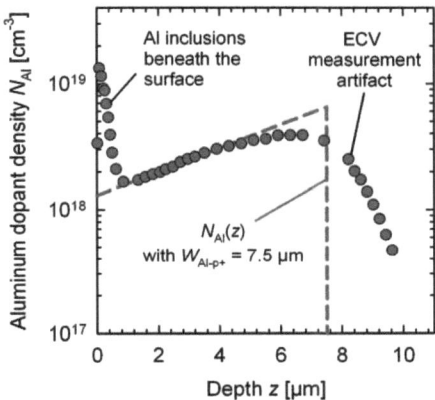

Figure 3.8.: *Aluminum doping profile of an Al-p^+ region (symbols) measured with the electrochemical capacitance voltage (ECV) method (taken from [11]). It is well described with the theoretical approximation in Eqn. 3.2, using $W_{Al-p^+} = 7.5 \mu m$. Note, that the peak at the front surface is not due to Al dopant atoms, but Al inclusions at the surface [11]. The broadening of the second peak can be attributed to artifacts of the ECV measurement [86].*

contacts due to a highly doped area at the interface of Si and metal. This in accordance with our arguments in chapter 1 and with literature data [33]. However, additionally we perform numerical simulations using PC1D [84] to verify the observed behaviour in Fig. 3.7. This will also verify our measurement technique for S_{cont} and allows predicting the contact recombination velocity S_{cont} from measurements of the mean Al-p^+ layer thickness.

The input parameters for the PC1D simulation are the base resistivity of 200Ωcm, the bulk lifetime in the Al-p^+ layer $\tau_{Bulk} = 130 ns$ [85] and a rear surface recombination velocity $S_{cell,rear}$. The measured Aluminum dopant density N_{Al} basically follows an exponential increase with depth z. It is increasing from $N_{min} = 10^{18} cm^{-3}$ at the sample surface to a maximum value of $N_{max} = 3 \times 10^{18} cm^{-3}$ at the interface of Si substrate and Al-p^+ layer, as shown exemplarily in Fig. 3.8. We use a depth dependent Al doping profile

$$N_{Al}(z) = N_{min} exp \left(\frac{z}{W_{Al-p^+}} ln \left(\frac{N_{max}}{N_{min}} \right) \right) \quad (3.2)$$

to describe the experimental values in Fig. 3.8. Here W_{Al-p^+} denotes the Al-p^+ layer thickness.

We observe a distinct difference between the experimental and theoretical curve close to the sample surface and at the end of the Al-p^+ layer. However the deviation at both positions was found to be an artifact. The Al concentration peak close to the surface was found to be the result of Al inclusions beneath the surface [11]. The relatively broad second concentration peak has been attributed to be a measurement artifact of ECV measurements [86]. This argument is supported by the sharp transition from Al-p^+ layer to the Silicon bulk in Fig. 3.5.

We evaluate the open circuit voltages V_{oc} of a solar cell, comprising an Al-p^+ layer of certain thickness W_{Al-p^+} and a rear surface recombination velocity of $10^7 cm/s$ (cell type one) and a solar cell without Al-p^+ region, but a certain $S_{cell,rear}$ (cell type two). For a given W_{Al-p^+} of cell type one we change $S_{cell,rear}$ of cell type two, until both cell types feature the same V_{oc}. In this case we have $S_{cell,rear} = S_{cont}$. This procedure determines S_{cont} as a function of W_{Al-p^+}. The results of the simulations are plotted in Fig. 3.7. They clearly confirm the experimental finding and verify therefore also the measurement method.

We find a rather inhomogenous thickness of the Al-p^+ region in Fig. 3.5. Due to the inhomogeneity and the nonlinear impact of W_{Al-p^+} on S_{cont} shown in Fig. 3.7, the evaluation of an arithmetic mean of the Al-p^+ thickness contains an uncertainty. Especially edge regions of the contact are found to comprise a thinner Al-p^+ layer. Hence, the evaluation is less precise for point contacts compared to line contacts, due to the larger circumference to contact area ratio. This might explain why point contacts feature a higher S_{cont}, compared to line contacts of a similarly thick Al-p^+ layer. A more detailed discussion of an inhomogeneity in the Al-p^+ layer can be found in Ref. [55].

3.4. Conclusion

In this chapter we presented measurements of the recombination at local Al-alloyed contacts on high- and low-resistivity p-type FZ Si, employing the method presented in chapter 2. The measured contact recombination has been found to depend on the geometry and size of the contacts. Very low reverse saturation current densities of $J_{0,cont} = 9 \times 10^2 fA/cm^2$ at the contacts on the $1.5\Omega cm$ resistivity material and $J_{0,cont} = 2 \times 10^4 fA/cm^2$ at the $200\Omega cm$ resistivity material demonstrate the significant improvement in contact recombination, employing local Al-alloyed line contacts.

Contact recombination velocities of line contacts are found to be several orders of magnitude lower than those of single-point contacts. When increasing the point contact size the point contact recombination decreases to similar values of those at line contacts. The assumption of reduced contact recombination due to the formation of a locally highly Al doped Al-p^+ layer beneath the contacts is proven, using SEM with secondary electron contrast. The thickness of the Al-p^+ region correlates with the measured contact recombination velocity. The agreement of the data with a PC1D simulation confirms this understanding. To achieve low contact recombination a more than $1\mu m$ thick Al-p^+ layer is necessary.

4. Understanding the local alloying of aluminum and silicon

In chapter 3 we demonstrated the enormous impact that the contact geometry has on the recombination at local Al alloyed contacts. We also identified that the change in contact recombination is the result of different Al-p^+ layer thicknesses. From this we conclude that the contact formation process is sensitive to the local contat geometry.

The impact of the local contact geometry on the contact formation process has been already described in various publications. Fischer for example worked on local alloying at point contact openings with a diameter down to $3\mu m$ [13]. He observed pyramidal shaped contacts without Al-p^+ layer. Uruena found that with increasing point contact size an Al-p^+ layer forms [14]. He introduced a qualitative model of the contact formation process based on the phase diagram of Al and Si [50]. Grasso et al. [15] measured the Al-p^+ layer thickness of point contacts as a function of the contact size.

However, a quantitative and general understanding of the contact formation process including line contacts was not yet developed. We therefore introduce in this chapter a quantitative model which is able to describe the contact formation at point and line contact geometries as a function of both: The local contact geometry and processing parameters, such as the firing temperature.

As the recombination at local Al alloyed contacts is basically dominated by the average Al-p^+ layer thickness W_{Al-p^+} we aim at describing W_{Al-p^+} with our model. We use here a basic analytic explanation of the process dependent on two parameters only. Our analytical approach is not intended to completely describe the contact formation. In fact, it simplifies the complex interdependencies by using physically justified assumptions. This work is already published in Ref. [46]

Another important phenomena often discussed in the context of local alloying is the formation of cavities or voids [13, 54–57]. However, in this work we focus on the Al-p^+ layer thickness only, as it is strongly correlated with the contact recombination.

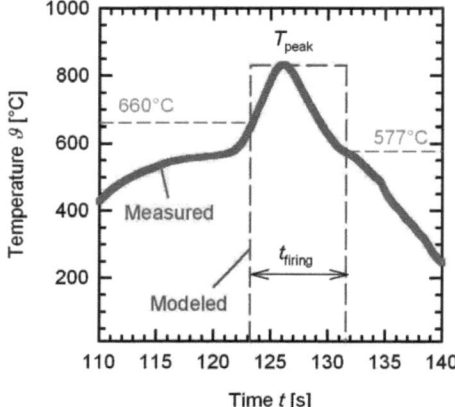

Figure 4.1.: *Sample temperature as a function of time t. The alloying process starts after the melting temperature of Al (660°C) is reached and ends after the sample is cooled down to the eutectic temperature of 577°C.*

4.1. Kinetic model of the local alloying process

In contrary to full area Al alloying the local contact formation cannot be treated as an equilibriuum process anymore. Due to the local structure of contacts the solution of Si and its transport in the alloy become tremendously important. We therefore analyze in the following the dynamics of the local alloying process.

4.1.1. The firing step

As introduced in chapter 1 the formation of the Al-p^+ layer takes place during a short firing step at the order of 10 seconds. During this short time the sample is heated above 800°C. The temperature of a sample during rapid thermal annealing is depicted in Fig. 4.1. After a preheat to 660°C the samples are heated up to the peak temperature of approximately $T_{peak} = 800°C$ in this case. After the peak temperature is reached the sample cools down to room temperature again.

The alloying process itself does take place for a very short time in this case of $t_{firing} = 8.4s$ only. It starts when the Al paste melts at 660°C and ends when the eutectic temperature of Al and Si of 577°C is reached. We therefore do only consider for our kinetic description the short step to very high temperatures. For a basic description of the alloying

4.1. Kinetic model of the local alloying process

process we assume a rectangular shaped temperature profile as sketched in Fig. 4.1.

4.1.2. Silicon concentration in time

A basic parameter to describe the dynamics of the contact formation process is the Si concentration

$$c_{Si}(t) = \frac{m_{Si}}{m_{Si} + m_{Al}} \tag{4.1}$$

in the Al melt. In the following we consider it as a function of time t. Here m_{Si} and m_{Al} denote the masses of Si and Al that contribute to the alloy in the melt. During alloying solid Si dissolves into the Al melt. As a consequence is the Si concentration increasing with time t.

For simplicity we assume c_{Si} to be spatially homogeneous in the alloy. This assumption keeps the model simple since it allows to decribe c_{Si} as a function of time t only which avoids coupled differential equations in space and time. However, measurements of the silicon distribution using energy dispersive X-ray spectroscopy (EDX) were pointing out that the silicon concentration in the alloy may vary laterally after firing [54]. Later on we will restrict the region where alloying occurs to a volume near the contact opening.

After the firing process the sample cools down as described in section 1.3.1. As the temperature decreases Si precipitates from the melt due to a limitation of its solubility in the alloy. Therefore the Si concentration in the melt c_{Si} decreases. The expelled Si atoms grow epitaxially onto the Si substrate and Al is incorporated into the lattice according to its solid solubility in Si which forms the Al-p^+ layer. When the liquid phase reaches the eutectic composition E of approximately 12% no more Si seggregates and the remaining liquid phase solidifies.

We therefore calculate the mass of recrystallized Si

$$m_{Si,Al-p+} = m_{Si,total} - m_{Si,eutectic} \tag{4.2}$$

as the difference of the total mass of dissolved Si $m_{Si,total}$ and the mass of Si in the eutectic layer $m_{Si,eutectic}$ remaining after the eutectic temperature is reached [82]. Together with the definition of the Silicon concentration c_{Si} we are able to calculate the mass Si defining the Al-p^+ layer

$$m_{Si,Al-p+} = m_{Al} \left[\frac{c_{Si}(t_{firing})}{1 - c_{Si}(t_{firing})} - \frac{E}{1 - E} \right] \tag{4.3}$$

as a function of the net mass of Al m_{Al} contributing to the alloying and the concentrations of Si in the alloy at the end of the firing process $c_{Si}(t_{firing})$ and at the end of the alloying E.

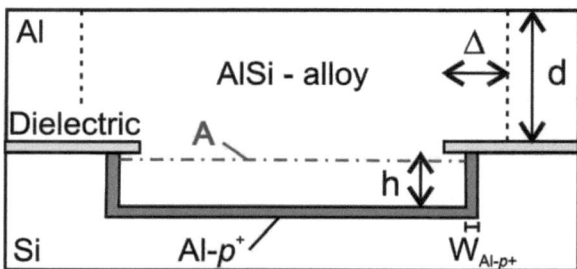

Figure 4.2.: *Sketch of a local Al alloyed contact (not to scale) illustrating the length Δ and the contact depth h.*

The dielectric layer between Al and Si acts as a solution barrier. Hence, the alloying process only takes place in the areas of the local contact openings (LCOs) where Al and Si are in direct contact. We denote this interface area of Al and Si with A. Using that we are able to calculate the thickness of the Al-p^+ layer

$$W_{Al-p+} = \frac{m_{Al}}{A\rho_{Si}} \left[\frac{c_{Si}(t_{firing})}{1 - c_{Si}(t_{firing})} - \frac{E}{1-E} \right] \tag{4.4}$$

where ρ_{Si} denotes the density of Si.

4.1.3. Al paste consumption

In Ref. [54] a restriction of the Si transport in the Al paste was found in the case of local contacts. Using EDX analysis and micrographs they found Si only near the interface areas A of Al and Si. This result implies that only a fraction of the total amount of Al paste actually contributes to the alloying process. It confirms earlier experiments by Huster et al. [87] who found evidence for a Si transport restriction in the case of a full area alloying process.

We consider the Si transport restriction during alloying by introducing the length Δ, which is illustrated in Fig. 4.2. It describes the region in the vicinity of the LCO to which Si is transported during firing. We therefore obtain for the mass of Al contributing to the alloy

$$m_{Al} = d\pi(r + \Delta)^2 \rho_{Al} \tag{4.5}$$

for local point contacts with radius r and

$$m_{Al} = d(a + 2\Delta)p\rho_{Al} \tag{4.6}$$

4.1. Kinetic model of the local alloying process

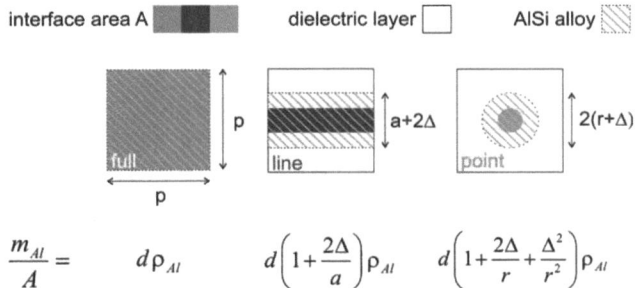

Figure 4.3.: Ratio m_{Al}/A illustrated for a full area, line and point contact. The length of the unit cell equals the contact pitch p.

for line contacts of width a. Here d denotes the thickness of the deposited Al layer and ρ_{Al} the density of the Al layer. In case of a thick Al paste there might be also a vertical transport limitation as described by Huster et al. [87]. Therefore a more general approach would be to define rather a volume of contributing Al instead of the length Δ. To keep our model simple, we further use horizontal transport restriction only described by Δ.

We now analyse the ratio m_{Al}/A (mass of Al per interface area between Al and Si) in Eqn. 4.4 as a function of the local contact geometry in Fig. 4.3. Here, we use the contact pitch p to describe the area of the unit cell p^2 which is equal to the interface area A when considering a full area contact. In the case of local contacts the interface area reads

$$A = \pi r^2 \qquad (4.7)$$

for local point contacts with radius r and

$$A = p \times a \qquad (4.8)$$

for line contacts of width a. As a result we obtain the lowest m_{Al}/A in the case of full area contacts translating to a thin Al-p^+ layer in Eqn. 4.4. With increasing length Δ the ratio m_{Al}/A increases. Considering point and line contacts the ratio m_{Al}/A decreases with increasing contact size. Thus, the amount of Al m_{Al}, which is contributing to the alloy as well as the interface area A of Al and Si are of major importance.

4.1.4. Dissolution of silicon in the melt

The dissolution of solid Si is one of the processes which determines the time dependence of the Si concentration in the melt c_{Si} significantly. The change of c_{Si} with respect to time

$$\frac{dc_{Si}}{dt} = k[F - c_{Si}] \tag{4.9}$$

is directly proportional to the difference of the equilibriuum concentration F and c_{Si}, where k is the constant of proportionality. Hence, the driving force for the solution process is high in case no Si is solved in the Al melt and the Si concentration c_{Si} increases fast. On the contrary in case the equilibriuum concentration $c_{Si} = F$ has been reached in the melt no further solid Si dissolves.

The solution of Eqn. 4.9

$$c_{Si}(t) = F\left(1 - e^{-kt}\right) \tag{4.10}$$

describes an exponentially saturating content of Si in the Al melt with respect to time.

We determine the constant of proportionality k by using the boundary condition $c_{Si} = 0$ at the beginning of the firing process. In this case we get the following approximation for the Si concentration

$$c_{Si} \sim \frac{m_{Si}}{m_{Al}} \tag{4.11}$$

from Eqn. 4.1. Note, that Al pastes can also contain Si [7,48] which would than change the boundary condition. The mass of dissolved Si

$$m_{Si} = h(t) \times A \times \rho_{Si} \tag{4.12}$$

however can be calculated by multiplying the volume of solved Si times the mass density of Si. Here h denotes the time dependent depth of the contact. From these arguments we conclude

$$\frac{dc_{Si}(t=0)}{dt} \sim \frac{1}{m_{Al}}\frac{dm_{Si}}{dt} = \frac{A\rho_{Si}}{m_{Al}}\frac{dh}{dt} \tag{4.13}$$

Hence, from 4.9 and 4.13 we finally calculate the constant of proportionality

$$k = \frac{A}{m_{Al}}\frac{\rho_{Si}v_{diss}}{F} \tag{4.14}$$

at $t = 0$. Here $v_{diss} = dh/dt$ describes the dissolution velocity of a Si layer per time for low Si concentrations $c_{Si} \sim 0$ in the alloy. Finally we are able to calculate the Al-p^+ layer thickness

$$W_{Al-p^+} = \frac{m_{Al}}{A\rho_{Si}}\left[\frac{c_{Si}(t_{firing})}{1 - c_{Si}(t_{firing})} - \frac{E}{1-E}\right] \tag{4.15}$$

from Eqn. 4.4 with

4.1. Kinetic model of the local alloying process

$$c_{Si}(t_{firing}) = F\left(1 - exp\left(-\frac{A}{m_{Al}}\frac{\rho_{Si}v_{diss}}{F}t_{firing}\right)\right) \tag{4.16}$$

by considering the respective ratio m_{Al}/A for the point or line contact geometry. In the case of a negative result in Eqn. 4.15 W_{Al-p^+} is zero.

Before we interpret our model in Eqn. 4.4 as a function of the processing conditions we determine in the following the two model parameters v_{diss} and Δ.

4.1.5. Experimental verification of the model

We experimentally verify the proposed model by comparing it to measured data. Additionally we determine the two model parameters v_{diss} the dissolution velocity of solid silicon during alloying and the length Δ describing the vicinity of the contact area from which additional Al is contributing to the contact area.

For this purpose we prepare local Al alloyed contacts on FZ silicon samples by local laser ablation of the dielectric layer, subsequent screen printing of an Al paste and firing of the sample. Experimental details are described in chapter 3.

We prepare point and line contacts with point contact radii 50 $\mu m < r < 290$ μm and line width 100 $\mu m < a < 280$ μm after firing. From time-dependent temperature measurements similar to the one shown in Fig. 4.1 we deduce a time $t_{firing} = 10s$ for the firing step. The peak firing temperature was $800°C$ defining the equilibriuum Si concentration F to be 0.27 according to the AlSi phase diagram. Considering a sphere packing of the Al paste particles we use $d = 26\mu m$ to calculate m_{Al} in 4.15 after measuring the Al paste thickness employing SEM images.

Following the method described in chapter 3 we determine a mean Al-p^+ layer thickness W_{Al-p^+} for the analysed contact geometries. The measured W_{Al-p^+} is shown in Fig. 4.4 as a function of the local contact structure size a and r after firing. We describe the measured data in Fig. 4.4 using the values for F, t_{firing} and d in Eqn. 4.15. From a least-square fit of Eqn. 4.15 to the experimental data we deduce the two parameters v_{diss} and Δ.

From the point and line contact data we obtain similar dissolution velocities of solid Si of $v_{diss,points} = 1.9\mu m/s$ and $v_{diss,lines} = 1.6\mu m/s$. To our knowledge no comparable data for this process exists. Dissolution velocities of Si during wet chemical etching at room temperature are at the order of $v_{diss} \sim 1\mu m/s$ [42] and therefore quite similar although at a 800K lower temperature. The length $\Delta_{points} = 120\mu m$ is larger for the point contacts compared to $\Delta_{lines} = 70\mu m$ for line contacts. The value of $70\mu m$ in the case of line contacts is supported by similar values determined from optical microscopy [54] studies.

The difference in Δ between point and line contacts is a consequence of the different dynamics of Si transport in the Al melt for point and line sources of Si. When the firing process begins and Si starts to dissolve into the Al melt the Si concentration in the melt

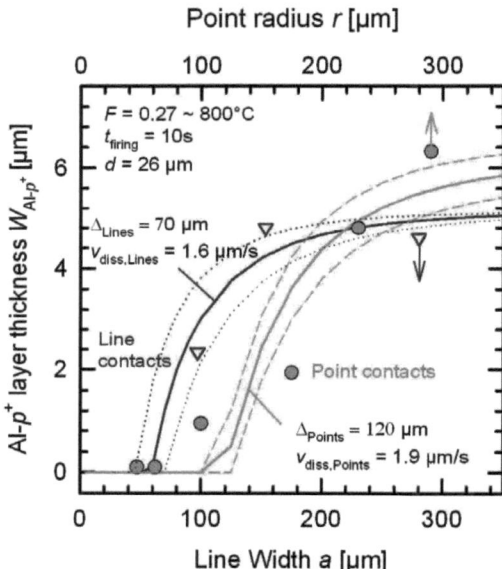

Figure 4.4.: *Measured Al-p$^+$ thickness at point (green circles) and line contacts (blue triangles down) and fit of our analytical model of contact formation to the measured data. The uncertainty in the determination of $v_{diss} \pm 0.1\mu m/s$ (dashed lines) and $\Delta \pm 15\mu m$ (dottel lines) is exemplarily demonstrated.*

is zero. Considering diffusion as the dominant mechaism of Si transport, we expect Si diffusion from the local contact source with Si concentration $c_{Si} = c_{Si,0}$ into the alloy. Analyzing the solution of the diffusion equation

$$\frac{\partial c_{Si}}{\partial t} = D \frac{\partial^2 c_{Si}}{\partial \mathbf{r}^2} \qquad (4.17)$$

we find now different solutions for point and line sources. Here $\mathbf{r}$ denotes the vector of location. In the case of point sources diffusing into a plane we have

$$c_{Si}(\mathbf{r},t) = \frac{c_{Si,0}}{4\pi Dt} \times exp\left(-\frac{r^2}{4Dt}\right) \qquad (4.18)$$

with D the diffusion coefficient and $c_{Si,0}$ the intial Si concentration. In the case of line sources diffusing into a plane we have a 1-dimensional problem with

$$c_{Si}(\mathbf{r},t) = \frac{c_{Si,0}}{\sqrt{4\pi Dt}} \times exp\left(-\frac{r^2}{4Dt}\right) \qquad (4.19)$$

For this reason we get at the same distance $\|\mathbf{r}\|$ of the contact a roughly twice as high Si concentration at point contacts compared to line contacts $c_{Si,points} = 2c_{Si,lines}$. From this argument we conclude that the Si is able to penetrate laterally much wider in the Al melt in the case of point contacts. This might explain the higher Δ_{points} compared to Δ_{lines}, as Δ was defined to describe a region in the vicinity of the contact to which Si is transported during firing.

The uncertainty in v_{diss} and Δ is exemplarily demonstrated in Fig. 4.4 by dotted and dashed lines. As shown for line contacts, a variation in Δ by $\pm 15\mu m$ exhibits a decreasing impact on the Al-p^+ layer thickness for an increasing structure size. On the contrary, a small variation in v_{diss} by $\pm 0.1\mu m/s$ shows an increasing impact on the Al-p^+ layer thickness with increasing structure size demonstrated for the point contact data. Hence, from a fit of Eqn. 4.15 to the experimental data Δ and v_{diss} can be determined independently.

4.2. Considering spacing of the contacts

In the previous section we demonstrated that our quantitative model reproduces the measured Al-p^+ layer thickness as a function of the point and line contact size. Now, we explain another effect first described by Urrejola et al. [47] and Rauer et al. [48]. They found an increasing thickness W_{Al-p^+} of the Al-p^+ layer with a decreasing pitch p of line and point contacts. The significant impact of the contact pitch p is shown in Fig. 4.5. With decreasing pitch the contact width after firing a_f and the contact depth decreases, whereas the Al-p^+ layer thickness W_{Al-p^+} increases.

In the frame of our model we explain this effect by a reduction of m_{Al} due to the small

Figure 4.5.: *SEM images of line contacts of width $a = 30\mu m$ before screen printing. The contact formation strongly depends on the contact pitch p. The micrographs in the lower part show the LCOs before screen printing.*

contact spacing p. This reduction then leads to a reduction of m_{Al}/A, which increases W_{Al-p+}.

4.2.1. Extension of the kinetic model

We extend our model to consider the impact of contact pitch p in the case of line contacts. We explain the observed effect by analyzing the transport of Si into the alloy during contact formation. In the previous section we found that the transport of Si into the Al melt is restricted to a volume near the interface area of Al and Si. For this purpose we introduced a length Δ, which describes the region in the vicinity of the contact area where the alloying process takes place.

Now, in the case of a small line contact pitch $p < 2\Delta + a$ the areas of neighbouring contacts where the alloying takes place overlap. As a result, the mass of Al contributing to the alloying decreases from $m_{Al} = d(a + 2\Delta)p\rho_{Al}$ when $p > a + 2\Delta$ to

$$m_{Al} = dp^2 \rho_{Al} \qquad (4.20)$$

in the case of $p < a + 2\Delta$. Here d denotes the Al paste thickness and ρ_{Al} the density of Al. As a result also the ratio m_{Al}/A decreases to

$$\frac{m_{Al}}{A} = d\rho_{Al}\frac{p}{a} \qquad (4.21)$$

4.2. Considering spacing of the contacts

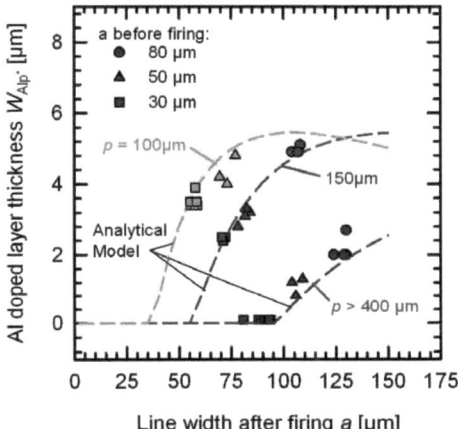

Figure 4.6.: Local Al-p^+ layer thickness W_{Al-p^+} as a function of line width a after firing (symbols). The analytical model in Eqn. 4.15 together with the extension in Eqn. 4.21 describes the impact of the contact pitch p without free fit parameter (lines).

when $p < a + 2\Delta$.

We apply this extension to experimental data in Fig. 4.6. Here W_{Al-p+} is measured as a function of the line width after firing a for contact pitches between $100\mu m < p < 4mm$. We observe in Fig. 4.6 that the Al-p^+ layer thickness is strongly dependent on both the contact pitch p as well as the line width a in this case measured after firing. In the case of large contact pitches $p > 400\mu m$ an Al-p^+ layer will only form for line widths after firing $a > 100\mu m$. In contrast, for a contact pitch of $p = 100\mu m$ an Al-p^+ layer will already form for line widths $a > 40\mu m$.

We now apply the extension of our kinetic model for contact formation to the data in Fig. 4.6. The thickness of the Al paste was $d = 35\mu m$. We insert the ratio m_{Al}/A determined from Eqn. 4.21 into the analytical model for the Al-p^+ layer thickness in Eqn. 4.15. This yields the dashed lines shown in Fig. 4.6 which are in excellent agreement to the experimental data. Note, the fit of our model does not include any additional free fit parameter.

Note, that we expect a similar extension to be applicable for point contacts. Rauer et al. [48] demonstrated that the Al-p^+ layer thickness of point contacts increases with decreasing contact size. However, our extension is so far only applicable to line contacts.

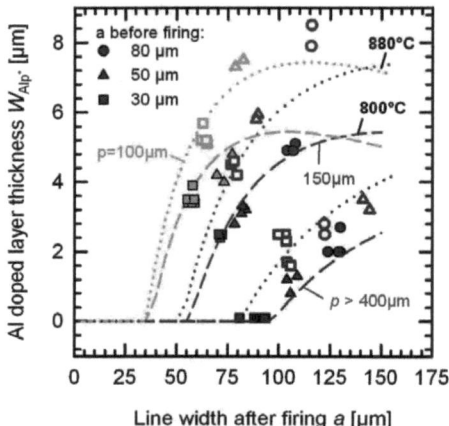

Figure 4.7.: *Al-p^+ layer thickness W_{Al-p^+} as a function of line contact width a after firing (symbols). Here we compare data from samples fired at 800°C (closed symbols) and 880°C (open symbols). We explain the measured data again with the model in Eqn. 4.15 and its extension in Eqn. 4.21 (800°C dashed lines; 880°C dotted lines).*

4.2.2. Impact of peak firing temperature

We further validate the kinetic model in Eqn. 4.15 and its extension in Eqn. 4.21. For the derivation of our model we assumed a rectangular shaped temperature profile (dashed line in Fig. 4.1). This approximation is however very basic. We therefore test our approach at two identical samples, fired at two different temperatures of 800°C and 880°C in Fig. 4.7.

Again, we measure the Al-p^+ layer thickness W_{Al-p^+} as a function of line contact width a after firing. To describe the measured data we use the same model parameters as in Fig. 4.6. However, instead of $F = 0.27$ which is the equilibriuum Si concentration at 800°C in the phase diagram of Al and Si we use $F = 0.34$ to describe the data at 880°C. Again we observe an excellement agreement of the measured data and our kinetic model in Eqn. 4.15.

4.3. Predictions of this model

Now we use our model (Eqn. 4.15) and the extension (Eqn. 4.21) for a general interpretation of local Al alloyed contact formation. For this purpose we analyse Eqn. 4.15 as a

4.3. Predictions of this model

Table 4.1.: *Overview of process parameters affecting contact formation in Eqn. 4.15*

Quantity	Description	Affected by
F	Si equlibriuum concentration	Firing temperature
m_{Al}	Amount of Al contributing to the melt	Paste thickness and contact geometry
A	interface area	contact geometry
t_{firing}	Time of firing step	Belt speed

function of the local contact geometry and the three main process parameters: The firing temperature ϑ_{peak}, the firing time t_{firing} and the paste thickness d. In Table 4.1 we link those process parameters to the corresponding quanitities used in Eqn. 4.15.

Increasing for example the peak firing temperature T_{peak} in Fig. 4.8 increases the maximum Al-p^+ layer thickness W_{Al-p^+}. A high peak firing temperature T_{peak} allows for a high equilibrium Si concentration F in Eqn. 4.15. For the parameter range chosen in Fig. 4.8 a high F therefore translates into a thick Al-p^+ layer. In the case of the small contact pitch of $p = 100\mu m$ we observe a decrease in W_{Al-p^+} for contact widths $a > 100\mu m$. For such geometrical arrangements of large contacts with little space in between the Si concentration at the end of the firing process approaches the equilibriuum Si concentration F and with decreasing ratio m_{Al}/A also the Al-p^+ layer thickness decreases.

Now we analyse the impact of the time of firing t_{firing} in Fig. 4.9. Here we observe a significant increase in W_{Al-p^+} with t_{firing} as well as a substantial change in the curve shape. Increasing the firing time from $t_{firing} = 10s$ to $15s$ shifts the optimum line width a in terms of W_{Al-p^+} to smaller line widths. Furthermore the smallest line width allowing for the formation of an Al-p^+ layer is smaller than $a = 50\mu m$ for $t_{firing} = 15s$ compared to $a = 75\mu m$ for $t_{firing} = 10s$ when considering large contact pitches $p > 400\mu m$.

Printing a thinner layer of Al paste on to the solar cells results in an increase of the Al-p^+ layer thickness in Fig. 4.10. In the case of large contact pitches $p > 400\mu m$ and a thickness of the Al-paste of $d = 60\mu m$ we even observe in Fig. 4.10 that no Al-p^+ layer will form for line widths smaller than $150\mu m$. In the case of $p = 100\mu m$ the optimum line width as well as the smallest line width allowing for Al-p^+ layer formation shift down to smaller line widths considering a thinner Al-p^+ layer.

From this analysis of the proposed model we are able to understand formation of local Al-alloyed contacts in a more general way. We saw how the main process parameters: local contact geometry, firing temperature ϑ_{peak}, firing time t_{firing} and paste thickness d affect the formation of the $Al - p^+$ layer. The ratio m_{Al}/A is mainly determined by the contact geometry and paste thickness. This ratio however has a large impact on the Si

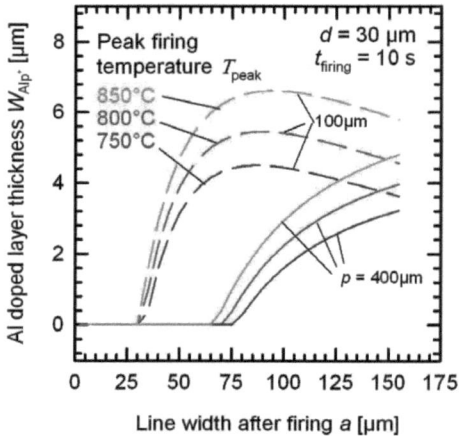

Figure 4.8.: Al-p^+ layer thickness as a function the line contact width a. Here the impact of the peak firing temperature is demonstrated by comparing $T_{peak} = 750°C$ (blue), $800°C$ (purple) and $850°C$ (green).

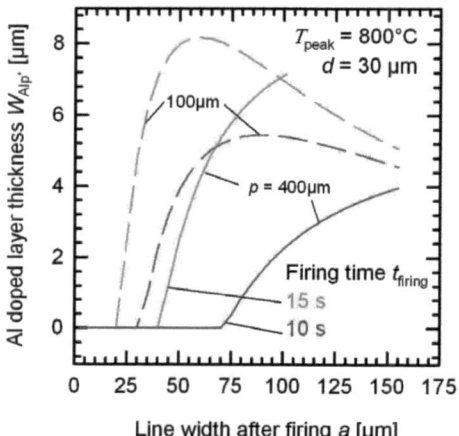

Figure 4.9.: Al-p^+ layer thickness W_{Al-p^+} as a function of the line contact width a. We compare a firing time t_{firing} of $15s$ (green) with $t_{firing} = 10s$ (purple).

4.4. Conclusion

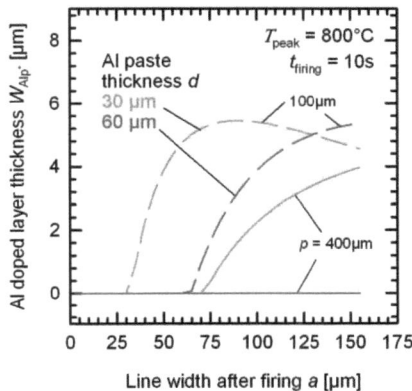

Figure 4.10.: Al-p^+ layer thickness W_{Al-p^+} as a function of the line contact width a. Here we analyze the impact of the Al paste thickness $d = 30\mu m$ (green) and $60\mu m$ (purple).

concentration at the end of the firing process as well as the overall Al-p^+ layer thickness in Eqn. 4.15. We find the following guideline for the impact of the processing parameters.

- The thickness d of the Al paste should be kept low to achieve a thick Al-p^+ layer.

- The thickness of the Al-p^+ layer shows an optimum as a function of the line width a. For small line widths the interface area A of Al and Si remains small and therefore also the Si concentration at the end of the firing process. However, in the case of wide line openings the ratio m_{Al}/A decreases which results in a decrease of W_{Al-p^+}.

- A high peak firing temperature ϑ_{peak} and a firing process with long firing time t_{firing} result in a thick Al-p^+ layer.

Note, however that any change in the firing conditions may affect the front contact formation or the surface morphology of the rear contact [88]. We therefore recommend to optimize paste thickness and the local contact geometry only.

4.4. Conclusion

Our quantitative model reproduces the measured impact of the point and line contact size on the Al-p^+ layer thickness. Our analytical approach is not intended to completely describe the contact formation. In fact, it simplifies the complex interdependencies by using physically justified assumptions. Following our explanation of the observed effects, the Al-p^+ layer thickness critically depends on the ratio m_{Al}/A. Thus, the amount of Al m_{Al} which is contributing to the alloy as well as the interface area A of Si and Al are of major importance.

This approach allows to determine the dissolution velocity of Si v_{diss} and the area where alloying takes place represented by the length Δ for a high temperature process with a duration of several seconds only. We determine $v_{diss,points} = 1.9 \mu m/s$ for point and $v_{diss,lines} = 1.6 \mu m/s$ for line contacts. The length $\Delta_{points} = 120 \mu m$ is larger for the point contacts compared to $\Delta_{lines} = 70 \mu m$ for line contacts. As the reason for this result we suppose the different symmetry of point and line contacts. This difference in symmetry influences the Si transport in the alloy when assuming diffusion as the relevant transport mechanism. In a recent publication [89] a similar but more detailed model was proposed where also diffusion is assumed to be the relevant transport mechanism.

An extension of the model considering the impact of the contact pitch was shown to describe experimental data very well. We explain the effect using a limitation in the area where alloying takes place. Hence, for small contact pitches these areas overlap and therefore the m_{Al} decreases.

5. Ablation of the dielectric layer

The alloying process for the formation of local Al alloyed contacts only occurs at the interface between aluminum and silicon. Removing the dielectric layer between aluminum and silicon before screen printing is therefore necessary for contact formation. Several technologies for the removal of the dielectric layer exist. The following processes were applied for the formation of local Al-p^+ layers in the past:

- **Al paste fire through**
  Here the Al paste is only locally printed [51, 52, 90]. Adopting the paste or the dielectric layer the paste will fire through the dielectric layer. Since it is printed only locally a local contact will form.

- **Etching paste**
  An etching paste based on phosphoric acid [7, 49] is locally printed on the dielectric layer [91]. During a thermal activation step the dielectric layer is etched away in the areas where the etching paste has been printed. Finally a cleaning step is necessary before printing of the Al paste.

- **Laser ablation**
  Before screen printing of the Al paste the dielectric layer is locally ablated using ns, ps or fs laser pulses. [44, 49, 53]

Most promising for industrial production in terms of process stability and cost is the laser ablation process. We therefore analyse this process further.

5.1. Comparison of ps and ns laser processes

For laser ablation of dielectric layers ns lasers are attractive due to their low cost. However, in comparison to ps lasers they demonstrated poorer performance in terms of recombination when applied to highly doped layers [92]. According to Ref. [92] and [78] a severe damage is introduced by ns laser pulses in the underlying Si during the ablation of a dielectric layer. In comparison ps laser pulses are known to ablate the dielectric layer almost without damage [73].

As these investigations have been performed on diffused samples the situation may differ significantly for Al alloyed contacts, where the laser treated area recieves a subsequent thermal treatment. In the following we therefore compare ps and ns laser processes in terms of recombination and analyse the contact formation process. We already published parts of this work in Ref. [44].

5.1.1. Contact recombination

We measure the contact recombination following the technique presented in chapter 2 and 3. For this purpose we prepare samples following the process sequence presented in chapter 3. However, we perform a different laser ablation process as described in the following:

Single-sided laser contact openings (LCOs) on planar surfaces are obtained by local laser ablation of the dielectric stack. We compare three different ablation processes at two laser wavelengths:

- 532nm, 10 ps long pulse length with a Gaussian beam profile (ps Laser) [92]
- 355nm, 20 ns long pulse length with Gaussian beam profile (ns Laser 1) and
- 532nm, 10 ns long pulse length with a flat-top beam profile (ns Laser 2) [93]

We realize openings of radius r and line contacts of line width a, by overlapping single openings of $25\mu m$ radius using the ps laser, $15\mu m$ radius in the case of ns Laser 1 and $85\mu m$ radius for ns Laser 2. The laser defines 16 contact geometries on the samples, each $2.5 \times 2.5 cm^2$ large. The geometries vary in contact pitch p with $p > 500\mu m$ or contact size (i.e. radius r or line width a). We use sample resistivities of $200\Omega cm$ and $1.5\Omega cm$.

In Fig. 3.4 the point contact data exhibits a decrease in S_{cont} with increasing contact size, independent of the material resistivity and the laser used. In contrast, the S_{cont} values of the line contacts are smaller and nearly independent of the line width in the range $80 < a < 300\mu m$ in Fig. 3.4. Comparing the contact recombination velocity for low resistivity $1.5\Omega cm$ material with those values obtained on $200\Omega cm$ material, a significantly higher recombination velocity (around one order of magnitude) is found for the low resistivity material. The lowest contact recombination velocity of $S_{cont} = 65 cm/s$ is observed for line contacts of width $a = 250\mu m$ processed with the ps Laser on $200\Omega cm$ material in Fig. 3.4. In contrast, the contact reverse saturation current density $J_{0,cont}$ is lower on low resistivity material. Values of $J_{0,cont}$ as low as $900 fA/cm^2$ are measured for line contacts on $1.5\Omega cm$ material in Fig. 3.4. These results obtained from Fig. 3.4 are in accordance to those in chapter 3 independent of the laser used for the ablation process.

Furthermore, the contact recombination depends on the laser used for the LCOs. Lower contact recombination is feasible with contacts opened with the ps Laser, compared to ns

5.1. Comparison of ps and ns laser processes

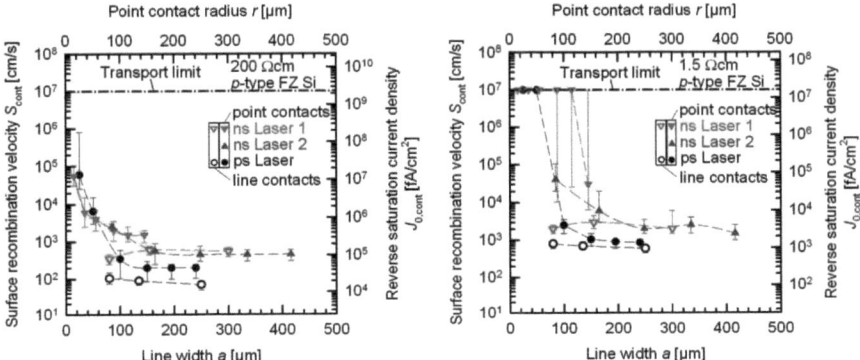

Figure 5.1.: *Contact recombination velocity S_{cont} and contact reverse saturation current density $J_{0,cont}$ at point contacts (closed symbols) as a function of the contact radius r and line contacts (open symbols) as a function of the line width a, measured on wafers of $200\Omega cm$ (left) and $1.5\Omega cm$ resistivity(right). The lines are guides to the eye. The thermal velocity equals the transport limit.*

Laser 1 and 2 in Fig. 3.4. The two ns laser processes are compareable in terms of S_{cont}. They demonstrate up to one order of magnitude higher contact recombination velocities S_{cont} compared to the ps laser process. However, this difference is less pronounced in the case of the $1.5\Omega cm$ material.

5.1.2. Al-p^+ layer thickness

In the previous section we observed a significant difference in the contact recombination velocity S_{cont} between the ns and ps laser processes. To give an explanation for this effect we compare micrographs and SEM images of ps Laser and ns Laser 1 line contacts in Fig. 5.2.

We observe in the SEM images in Fig. 5.2 a thinner Al-p^+ layer at the ns Laser 1 contacts compared to the ps laser contacts. Furthermore the contact formation is rather inhomogeneous with valleys and peaks of the Al-p^+ layer. Especially in the peak and valley regions the Al-p^+ layer thickness is strongly reduced.

From the SEM images we determine W_{Al-p+} as described in chapter 3 by measuring the cross sectional area of the Al-p^+ layer and dividing it by the contact width. We measure a mean Al-p^+ layer thickness of $W_{Al-p+} = 1.4\mu m$ at ns Laser 1 contact and $W_{Al-p+} = 4.8\mu m$ at the ps Laser contact. We consider the reduced thickness of the Al-p^+ layer as the cause for an increased contact recombination according to the correlation of S_{cont} and W_{Al-p+}

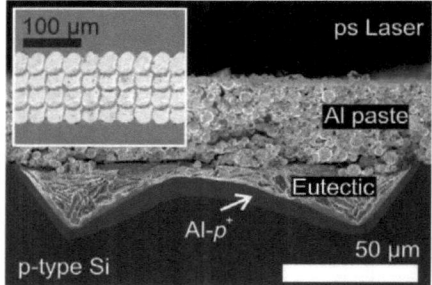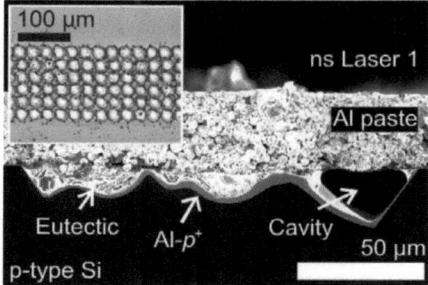

Figure 5.2.: *SEM image of a ps laser (left) and ns laser 1 (right) line contact of line width $a = 150\mu m$. In the upper left, a micrograph of the contact before screen printing demonstrates the different quality of the ps laser and ns laser 1 process.*

found in chapter 3.

The optical microscopic images in the upper left corner in Fig. 5.2 may give an explanation for the reduced Al-p^+ layer thickness at the ns Laser contacts. We find low quality LCOs with no overlap at the ns Laser 1 line contact. Hence, the alloying process takes place independently at each single point contact. Remains of the dielectric layer probably prevent Si to dissolve in the Al melt between the single LCOs. Additionally cavities form, which are found to result in a thinner local Al-p^+ layer. In contrast, at the ps Laser contact the alloying process takes place homogeneously over the whole laser ablated area. As a result a continuous Al-p^+ layer forms due to high quality and well overlapping LCOs.

In contrast to ps laser pulses, the relatively long ns laser pulses will affect the underlying silicon through heat dissipation in a depth of several μm [92] and defects are introduced in the Si bulk. However, while firing the sample, the Si dissolves into the Al melt. During cooling the liquid Si precipitates and grows defect free onto the Si substrate. A comparison of internal quantum efficiency (IQE) measurements of PERC cells comprising local Al alloyed contacts opened with UV ns laser and etching paste [53] demonstrated the general applicability of ns Lasers for local Al alloyed contacts. We therefore study ns laser ablation further in the next section.

Furthermore we find Si particles around the LCOs in the micrograph of the ns laser 1 contact. These particles increase the roughness of the surface. As a result of the surface roughness the Al melt may not be able to sufficiently wet the surface during contact formation [13] leading to thin Al-p^+ layers. We believe this is the main reason for the ineffective formation of the Al-p^+ layer. Therefore we introduce an additional KOH etching step in the next section to remove those particles.

5.2. Introducing an additional KOH etch

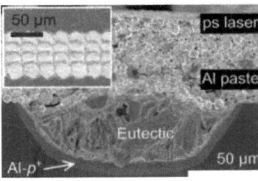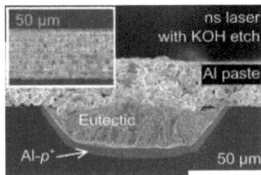

Figure 5.3.: *SEM images of a ps laser line contact (left), a ns laser line contact (middle) and a ns laser line contact with KOH etch (right). The insets show micrographs of the contacts before screen printing of the Al paste.*

5.2. Introducing an additional KOH etch

From the results in the previous section we developed the idea to combine the ns laser process with a subsequent KOH etching step. This etching step will provide a clean and smooth surface in the areas of LCOs. The basic idea is that remains of molten Si increase the surface roughness which makes wetting of the surface by Al paste impossible. Insufficient surface wetting might also explain the formation of cavities. This idea is in coincidence with the observations of Fischer [13]. He assumed that the cavities observed at the pyramid tips were due to insufficient surface wetting in the case of very small contacts.

5.2.1. Process optimization

For the ns laser ablation process we use ns laser 1, which has been introduced in section 5.1. However, by applying an improved optical setup we achieve with the same laser a focus of only $6\mu m$ diameter. This small laser focus is advantageous when aiming at very small structure sizes of our contacts. After ns laser ablation we apply an etch of 50 % KOH at $70°C$.

In Fig. 5.3 we compare SEM images and micrographs of a ps laser process (left), the ns laser process without KOH etch (middle) and the ns laser process with KOH etch. We observe no difference between the ps and ns laser process with KOH etch in the SEM images in Fig. 5.3. This demonstrates the excellent contact formation using a ns laser and KOH etch. We observe in the micrograph of the ns laser 1 process after KOH etch remains of the dielectric layer in between LCOs. However, the excellent contact formation applying the KOH step also proves that these remains do not necessarily hamper the contact formation.

Without the KOH etch step the contact formation is rather inhomogeneous at the ns laser contacts. The Al-p^+ layer thickness is even further reduced compared to the ns laser 1 process with the larger laser beam focus in Fig. 5.2. We explain this effect with an increase in surface roughness due to the high amount of valleys in the Si material as a

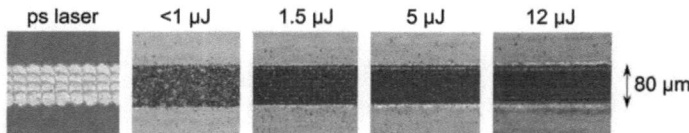

Figure 5.4.: *Micrographs of line openings of width $a = 80\mu m$. The dielectric layer was ablated using a ps laser and a ns laser. The ns laser pulse energy was varied between 1 and $12\mu J$.*

result of the ns laser ablation using the $6\mu m$ beam diameter.

After we demonstrated the general applicability of our process in Fig. 5.3 we first outline the optimization of our etch process.

Laser pulse energy

To implement the ns laser ablation with a subsequent etch in KOH solution, we start with analysing the impact of the laser pulse energy on the laser ablation process. For this purpose we apply line LCOs with a width of $80\mu m$ using a ps laser and the ns laser 1 with the $6\mu m$ laser focus. Experimental details of the sample preparation are described in chapter 3. For the ns laser ablation we use pulse energies between 1 and 12 μJ per pulse.

Micrographs of the samples after laser ablation shown in Fig. 5.4 demonstrate the high quality laser ablation of the ps laser process. Considering the ns laser ablation process we observe an increasing deformation of the sample surface with increasing pulse energy. To keep the impact of the ns laser processing as low as possible and to guarantee a stable laser pulse energy we decided to use a laser pulse energy of 1.5 μJ in the following investigations.

KOH etching time

In the next step we applied a KOH etch after the laser ablation. For solar cell application we need to consider not to deteriorate the passivation quality of the dielectric layer. We therefore chose an etching bath of 50 % KOH with a relatively low temperature of $70°C$. A similar solution has been already used by Mangersnes et al. [78] to remove ns laser damage after ablating a similar dielectric layer.

The KOH solution will however not only remove Si particles distributed by the ablation process, but will also etch the solid Si. With increasing etching time the more Si will dissolute and also the process cost will increase. For this purpose we study the contact formation as a function of the etching time to find the shortest possible etching time in Fig. 5.5. We determine the corresponding etch depth by gravimetry.

The etching time has been variied between 5 min and 30 min, which resulted in etch

5.2. Introducing an additional KOH etch

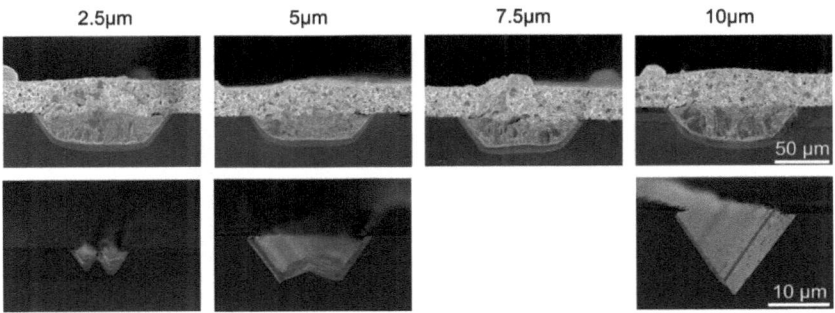

Figure 5.5.: *SEM images of ns laser contacts with KOH etch. The etching time has been variied between 5 min and 30 min, which resulted in etch depths between 2.5 and 10 µm. The images show contacts after contact formation (upper row) and prior to screen printing of the Al paste (lower row). The images demonstrate how solid Si is solved during etching with KOH which has only minor impact on the contact formation.*

depths between 2.5 and 10 μm. By comparing the SEM images of the contacts (upper row) in Fig. 5.5 we observe no significant difference for the different etching times. In the lower row SEM images demonstrate the contact shape before screen printing.

Due to the KOH etch pyramids form as a result of the preferred etching in (111) crystal orientation [13, 14]. With increasing etch time these pyramids enlarge and grow toegther in the case of very long etch times. However this has no impact on the contact formation process. We therefore use in the following an etch time of 5 mins only which equals an etch of 2.5μm Si. However, also shorter etching times might be applicable.

5.2.2. Comparison of ps and ns laser process with subsequent KOH etch

In the previous section we found that ns laser ablation processes are suitable for application to local Al alloyed contacts when combined with a KOH etch after laser treatment. This allows to use the ns laser process as an alternative to the ps laser process. Therefore we now evaluate the ns laser process in combination with KOH etch in terms of contact recombination by comparison to the ps laser process.

We prepare samples of 1.5Ωcm resistivity passivated on both sides following the procedure described in section 3.1. By overlapping of single LCOs using a ps and a ns laser we create line openings on the rear side of those samples. We apply lines of different pitch with a line width $a = 80\mu m$. First, the ns laser process is applied with a subsequent etch

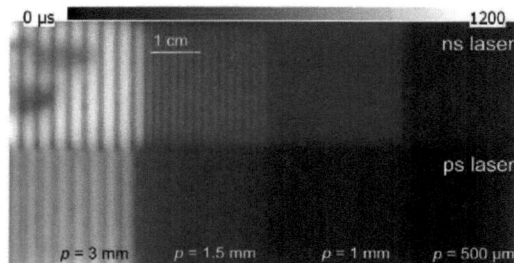

Figure 5.6.: *Dynamic ILM lifetime image of a $1.5\Omega cm$ sample with local line contacts of width $a = 80\mu m$ but different pitch p. Applying a ns laser process with KOH etch ($S_{cont} = 10^3 cm/s$) and a ps laser process ($S_{cont} = 3 \times 10^3 cm/s$) we find an improvement in the contact recombination velocity S_{cont} applying the ns laser process with KOH etch.*

in KOH. Then the ps laser process is applied to the remaining areas of the wafer.

We now measure measure the recombination properties of the ps and ns laser line openings. For this purpose we conduct dynamically calibrated steady-state ILM measurements in Fig. 5.6 as explained in chapter 2. Since equal contact geometries but different laser processes are side by side in Fig. 5.6 already a qualitative comparison of both processes can be done. Surprisingly we find higher lifetime values in the areas processed with the ns laser and additional KOH etch. This impressive result already confirms a lower contact recombination at the ns laser processed contacts, compared to those processed with the ps laser.

We evaluate the contact recombination velocity S_{cont} from the lifetime measurement in Fig. 5.6 using the procedure described in chapter 3. The result confirms that the contact recombination at the ns laser contacts with $S_{cont} = 10^3 cm/s$ is lower compared to the one at ps laser contacts with $S_{cont} = 3 \times 10^3 cm/s$. Note that the contact recombination of the ps laser contact is slightly higher compared to the results obtained in chapter 3. This may be the result of a difference in Al paste or its thickness in both experiments.

We attribute the reduction in contact recombination at ns laser processed contacts with additional KOH etch to a slight increase in the Al-p^+ layer thickness W_{Al-p+}. We measure in Fig. 5.3 at the ps laser processed contacts $W_{Al-p+} = 2.3\mu m$. However, in the case of a ns laser line opening with additional KOH etch we find a thicker Al-p^+ layer with $W_{Al-p+} = 3.4\mu m$.

5.3. Conclusion

In this chapter we compare two laser ablation processes with respect to their impact on the formation of local Al-alloyed contacts. The ps laser process demonstrated approximately one order of magnitude lower contact recombination velocities, compared to two ns laser processes. We identified the thinner Al-p^+ layer at the ns laser contacts as the origin of this effect. The thin Al-p^+ layer is the result of an inhomogenous contact formation with valleys and peaks of the Al-p^+ layer. We assume that the high surface roughness due to remains of molten Si after the laser ablation process hampers surface wetting of the alloy causing the inhomgenous contact formation.

We therefore introduce a KOH etch with 50 % KOH at 70°C after the ns laser ablation process to reduce the surface roughness. This process sequence delivers a good contact formation in a wide range of ns laser pulse energies and KOH etching times. The general applicability of this process has also been shown by Du et al. [94]. They applied a 10 % KOH subsequent to ps and ns laser ablation. Performing the additional KOH etch they could significantly improve the efficiency of solar cells processed with the ns laser. We were able to demonstrate a reduction in contact recombination velocity using the ns laser process with KOH etch resulting in $S_{cont} = 10^3 cm/s$ compared to the ps laser process resulting in $S_{cont} = 3 \times 10^3 cm/s$.

6. Simulation of silicon solar cells with local Aluminum alloyed base contacts

The performance optimization of solar cells comprising a locally contacted rear surface is a trade-off between rear surface recombination losses and resistive losses [21–23]. Considering local Al alloyed contacts formed by screen printing an additional challenge arises from the significant impact of the processing conditions on the contact properties which were described in the previous chapters.

In chapter 4 we developed an analytical model to describe the Al-p^+ layer thickness as a function of the processing conditions, i.e. contact geometry and firing conditions. Additionally we found in chapter 5 a significant impact of the laser process used for the local contact opening on the contact recombination. The specific contact resistance ρ_c of local Al-alloyed contacts measured on solar cells was found to be large independent of the actual contact geometry [95, 96].

The aim of this chapter is to integrate these properties of local Al alloyed contacts into a device simulation. We therefore developed an analytical spreadsheet calculator [97] which is able to find the optimum contact geometry based on experimental parameterizations and analytical models. For this purpose we extended an optimization tool for the rear contact geometry of solar cells introduced by Wolf et al. [20].

An analytical model does not consider the complete physics in a solar cell. Using numerical simulations solar cells with local Al alloyed contacts were simulated in the past [55, 98, 99]. However, in the frame of the multiple interdependencies of the processing parameters we think an easy to use and fast analytical simulation based on experimentally verified input parameters is a more suitable solution for practical applications in the case of local Al alloyed contacts.

6.1. Device Model

Instead of sophisticated numerical simulations we apply for our simulations an analytical model, which is based on the approach of Wolf et al. [20]. This model employs a description of the solar cell current-voltage characteristics with the 2-diode model, where the current density

80 Chapter 6. Simulation of silicon solar cells with local Aluminum alloyed base contacts

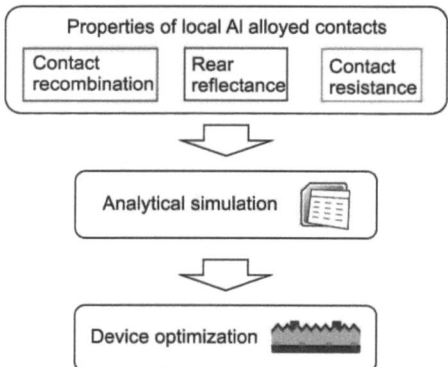

Figure 6.1.: *Schematic of the device optimization process. Models and parameterizations valid for Al-alloyed base contacts are employed in an analytical spreadsheet calculator.*

$$J = -J_{Ph} + J_{01}\left[\exp\left(\frac{V - JR_s}{V_{th}}\right) - 1\right] + J_{02}\left[\exp\left(\frac{V - JR_s}{2V_{th}}\right) - 1\right] + \frac{V - JR_s}{R_p} \quad (6.1)$$

is expressed as a function of the junction voltage V and the thermal voltage $V_{th} = 25.73mV$ at a temperature of $25°C$. Here J_{Ph} is the photogenerated current density; J_{01} and J_{02} are the first and second diode reverse saturation current densities. R_s and R_p denote the series and parallel resistance of the solar cell.

The 2-diode model is implemented in an analytical spreadsheet calculator. Since Eqn. 6.1 is an implicit function of the current density J, an iteration loop is used to calculate the open circuit voltage V_{OC}, the short circuit current density J_{SC} and the maximum power point (J_{mpp}, V_{mpp}). From these parameters fill factor FF and power conversion efficiency η are determined.

The schematic in Fig. 6.1 illustrates our approach. The parameters J_{Ph}, J_{01}, J_{02}, R_s, and R_p in the 2-diode model are calculated from a set of analytical models and experimentally verified parameterizations according to Ref. [20]. We only sketch here the main calculations which are performed in this model. To determine recombination in the locally contacted base the Fischer model [17] is used, which was introduced in chapter 2. Together with the bulk diffusion length L the effective diffusion lenght L_{eff} in the base follows. From this calculation J_{01} and J_{Ph} are determined dependent on the local contact geometry and the optical properties of the structure. For the calculation of J_{Ph} a parameterization of

6.1. Device Model

Fischer [17] is used, where a measured reflection spectrum is used to calculate the generated current density J_{gen}. We consider the impact of the local contact arrangement also for the calculation of the base series resistance, using analytical approximations for point [66] and line contact [67] layouts.

As an extension to the original model in Ref. [20] we consider the impact of the processing conditions to the local Al alloyed contact formation by implementing calculations of the contact resistance R_C and contact recombination velocity S_{cont}. Furthermore we consider the contact geometry dependent rear reflectance and consider its impact on the photogenerated current density J_{Ph}. We explain these extensions in more detail in the following sections.

6.1.1. Contact recombination

In chapter 3 we found a clear correlation between the contact recombination velocity S_{cont} and the mean thickness of the Al-p^+ layer W_{Al-p^+}. This allows us to calculate S_{cont} from W_{Al-p^+} using our kinetic model of the contact formation process which was presented in chapter 4.

We determined S_{cont} as a function of W_{Al-p^+} in chapter 3, however in a tedious procedure using numerical simulations with PC1D. Since we are aiming for an analytic simulation of solar cells we use here a different approach. We calculate in our simulations the contact recombination velocity

$$S_{cont} = S_{min} \coth\left(\frac{W_{Al-p^+}}{L_{Al-p^+}}\right) \qquad (6.2)$$

of local Al-alloyed contacts with a simple analytic equation from Ref. [33]. Here L_{Al-p^+} is the diffusion length of minority charge carriers in the Al-p^+ layer. Note, that Eqn. 6.2 has been derived for the case of homogenous doping and homogenous material properties. Godlewski et al. [33] calculated S_{cont} from material properties, such as doping or diffusion coefficients. We rather use the parameter S_{min} in Eqn. 6.2 as proportionality factor instead of the material properties. This approach allows

- neglecting the inhomogenous doping profile of the Al-p^+ which was discussed in chapter 1 and

- considering the impact of the laser process on contact recombination, which we analysed in chapter 5.

As a result S_{min} defines the lowest possible S_{cont} value in Eqn. 6.2. We determine S_{min} experimentally by measuring the contact recombination velocity S_{cont} as a function of the Al-p^+ layer thickness employing the procedure presented in chapter 3. S_{min} follows in Fig.

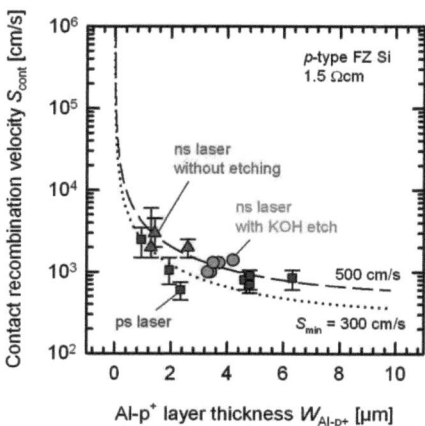

Figure 6.2.: *Contact recombination velocity S_{cont} as a function of Al-p$^+$ layer thickness W_{Al-p^+}. We determine for the ps laser process (blue squares) $S_{min,ps} = 300 cm/s$ and for the ns laser process without KOH etch (pink triangle up) $S_{min,ns} = 500 cm/s$ in Eqn. 6.2. The ns laser process with KOH etch (green circles) has been performed in a different experiment and is therefore excluded from the analysis of S_{min}*

6.2 from a fit of Eqn. 6.2 to the experimental data. Here the ps laser data is taken from chapter 3 and the ns laser data is taken from chapter 5.

To account for the significant impact of the laser ablation process on S_{cont}, we determine S_{min} for different laser ablation processes. For this purpose we prepare samples on $1.5\Omega cm$ FZ material using the following three laser ablation processes:

- ps laser ablation
- ns laser ablation
- ns laser ablation with subsequent KOH dip as presented in chapter 5

From the experimental data in Fig. 6.2 we determine $S_{min,ns} = 500 cm/s$ in the case of the ns laser process without additional KOH etch and $S_{min,ps} = 300 cm/s$ for the ps laser process. The data obtained for the ns laser ablation process with additional KOH etch can be described by either $S_{min,ps} = 300 cm/s$ or $S_{min,ns} = 500 cm/s$.

Note, that the samples of the ns laser process with KOH etch were fabricated in a different experiment than the samples of the ps laser and ns laser process without KOH

6.1. Device Model

etch. This may explain that in Fig. 6.2 the contact recombination is higher at ns laser contact with KOH etch compared to the ps laser process. For compareable contacts we found in chapter 5 a reduced contact recombination at the ns laser processed contacts with additional KOH etch compared to the ps laser processed contacts.

6.1.2. Base series resistance

The series resistance of screen-printed solar cells with local Al-alloyed contacts has been analyzed [95] for line contact arrangements of varying pitch p and line width a. These results demonstrated a remarkably high specific contact resistance $\rho_C = 55 m\Omega cm^2$ of local Al-alloyed contacts.

Urrejola et al. [100] found a significantly lower specific contact resistance of $9-17 m\Omega cm^2$ dependent on the local contact geometry. However, they applied local screen-printing of Al paste which may affect the contact formation. Other publications [96] confirmed the results obtained by Gatz et al. [95].

Using $\rho_C = 55 m\Omega cm^2$ together with an analytical model for the geometry dependent spreading resistance R_{spread} in the base for point [66] and line contacts [67] we are able to calculate the base series resistance

$$R_{s,base} = R_{spread}(f) + \frac{\rho_C}{f}. \tag{6.3}$$

Here, the rear metallization fraction f is for line contacts calculated as $f = a/p$ and for point contacts as $f = \pi r^2 / p^2$, with r denoting the point radius.

6.1.3. Optics

We measure the reflectance R of screen-printed Si solar cells as a function of the rear metallization fraction f. The results shown in Fig. 6.3 demonstrate the impact of f in the long wavelength range, where the optical properties of the cell rear dominate. Analyzing $R(\lambda)$ with a model introduced by Brendel et al. [101] we extract values for the rear reflectance R_b and the lambertian factor Λ as shown in the inset of Fig. 6.3.

Based on these parameters we calculate the generation profile $g(\lambda, z)$ as a function of the depth z. The cumulated generated current density

$$J_{gen}(z) = q \int_0^z \int \Phi_{AM1.5}(\lambda) g(z, \lambda) d\lambda dx \tag{6.4}$$

is then calculated considering an illuminating photon flux $\Phi_{AM1.5}$ corresponding to the AM1.5 spectrum. Aiming for a parameterization of J_{gen} as a function of the rear metallization fraction f, we approximate the obtained cumulated generated current density by

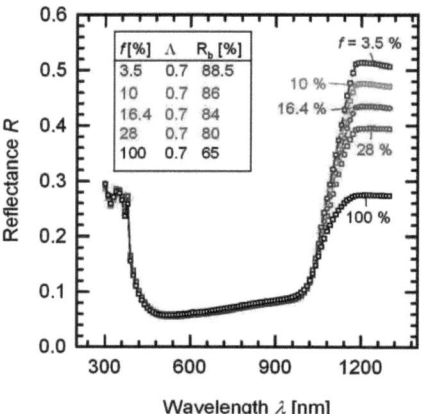

Figure 6.3.: *Reflectance R measured at fully screen printed Si solar cells with rear line contacts. For each metallization fraction f we determine the lambertian factor Λ and the rear reflectance R_b.*

$$J_{gen}(z) = (1-M)[J_{gen,front} + J_{gen,exp1}\left(1-e^{-z/L_1}\right) + J_{gen,exp2}\left(1-e^{-z/L_2}\right) + J_{gen,hom}\frac{z}{W}] \quad (6.5)$$

using a parameterization introduced by Fischer et al. [17]. This parameterization is a linear combination of four terms considering generation at the front surface (z=0) with $J_{gen,front}$. With $J_{gen,exp,1}$ and $J_{gen,exp,2}$ an exponentially decreasing generation is accounted for with an absorption length $L_1 = 4\mu m$ and $L_2 = 25\mu m$, respectively. The parameter $J_{gen,hom}$ describes homogenous generation due to weakly absorbed near infrared light. The factor $(1-M)$ accounts for shading by the front metallization grid.

The applicability of this approach is demonstrated in Fig. 6.4. Here $J_{gen,front}$, $J_{gen,exp1}$, $J_{gen,exp2}$ and $J_{gen,hom}$ are chosen to describe the cumulated generation current density of the solar cell with $f = 3.5\%$ in Fig. 6.3.

Independent of the rear metallization fraction f we obtain $J_{gen,front} = 7.5 mA/cm^2$, $J_{gen,exp,1} = 18.2 mA/cm^2$, $J_{gen,exp,2} = 10.2 mA/cm^2$ for each of the solar cells. Only the homogenous generation described by $J_{gen,hom}$ depends on the rear contact design as shown in Fig. 6.5. We describe $J_{gen,hom}(f)$ using a linear approach:

$$J_{gen,hom} = -\frac{3}{4}mA/cm^2 \times f + J_{gen,hom,f=0} \quad (6.6)$$

6.1. Device Model

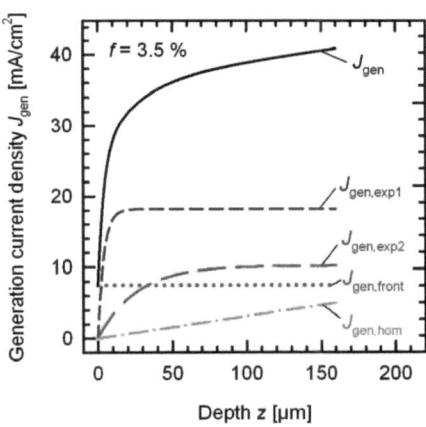

Figure 6.4.: *Generated current density J_{gen} as a function of the depth z. The solid line represents the cumulated generation current calculated from the reflectance curve of the cell with $f = 3.5\%$ shown in Fig. 6.3. Using Fischers parameterization we are able to describe $J_{gen}(z)$ using $J_{gen,front}$, $J_{gen,exp1}$, $J_{gen,exp2}$ and $J_{gen,hom}$.*

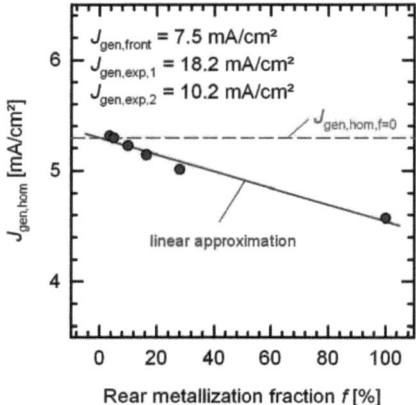

Figure 6.5.: *The parameter $J_{gen,hom}$ describing homogenous generation of free charge carriers over the sample thickness is a linear function of the rear metallization fraction f in Eqn. 6.6. The data is calculated from the reflectance spectra in Fig. 6.3*

Here $J_{gen,hom,f=0}$ denotes the homogenous generation current density in the case when no rear contacts would have been applied. The fact that $J_{gen,front}$, $J_{gen,exp,1}$ and $J_{gen,exp,2}$ are independent of f can be understood from the analysis of the reflectance data shown in Fig. 6.3. Only the long wavelength reflectance is affected by the rear metallization fraction f. Since long wavelength light is weakly absorbed in silicon, it results in a homogeneous generation. Using the parameters obtained by describing $J_{gen}(z)$, we calculate J_{Ph} according to Ref. [20].

6.2. Simulation results

In the previous section we explained our simulation model and the extensions necessary for the simulation of Aluminum alloyed contacts. Now, we employ our model to study the application of Aluminum alloyed contacts to solar cells. For this reason we study in the following the impact of parameters which influence contact recombination on solar cell performance. The impact of the optical properties and contact resistance is further analyzed in the appendix chapter B.

6.2. Simulation results

Table 6.1.: *Input parameters of the device model*

Parameter	Value	Description
J_{0e}	$250 fA/cm^2$	Front emitter saturation current density
J_{02}	$8 nA/cm^2$	2nd diode reverse saturation current density
$R_{s,front}$	$0.55 \Omega cm^2$	Lumped front series resistance
R_p	$10 k\Omega cm^2$	Parallel resistance
S_{pass}	$30 cm/s$	Rear surface recombination velocity in the passivated regions
R_{sh}	$60 \Omega/\square$	Emitter sheet resistance
γ_{inj}	1	Injection factor yielding the SRV at short circuit conditions
M	5%	Fraction of front surface shaded by contact grid
$\Delta J_{sc,emitter}$	$0.5 mA/cm^2$	Short circuit current density loss due to emitter recombination

6.2.1. Comparison with literature cell data

For a verification of our simulations we compare it to solar cell efficiencies η and fill factors FF, which have been recently reported in the literature [95]. For this purpose we perform simulations using device properties similar to the configuration reported in Ref. [95]. We simulate Czochralski (Cz) grown Si as base material with a thickness of $160 \mu m$ and a base resistivity of $2\Omega cm$. The bulk lifetime of $\tau_{Bulk} = 800 \mu s$ is related to the state after deactivation of the B-O defect [102,103]. We assume a thickness of the Al paste of $d = 35\mu m$ and a firing step at a temperature of $800°C$ with a length of $10s$. This defines the Al-p^+ layer thickness in our kinetic model of contact formation (see chapter 4). The rear contacts in Ref. [95] and our simulations are arranged in lines of width $a = 125\mu m$. For the ablation of the dielectric layer we considered the ps laser process unless otherwise stated defining $S_{min} = 300 cm/s$. The most relevant input parameters of the simulation are summarized in Table I.

As shown in Fig. 6.6, all solar cell efficiencies and fill factors of Ref. [95] are reported as a function of the line contact spacing p. Since no uncertainty was reported we presume an uncertainty in the efficiency η of $0.3\%_{abs}$ and in fill factor FF of $0.5\%_{abs}$ in accordance to typical uncertainty ranges given in calibration sheets.

The impact of the contact pitch p is well described by the simulations as can be seen from Fig. 6.6. The decrease in efficiency for large contact spacings is a result of the decrease in FF. The short circuit current density J_{sc} and open circuit voltage V_{OC} (not shown) are saturating for $p > 1mm$. The decrease in FF is due to further lateral transport of

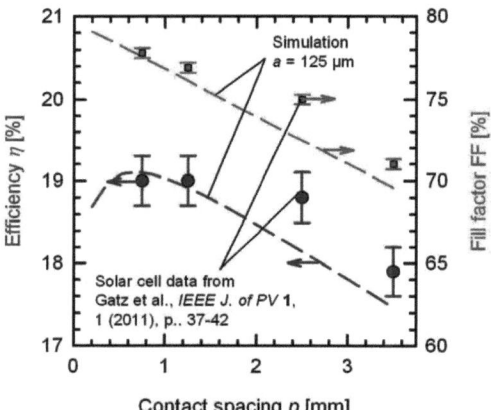

Figure 6.6.: *Simulated power conversion efficiency η (blue dashed line) and fill factor FF (red dashed line) using the parameters in table 6.1 compared with literature cell efficiencies (blue circles) and fill factors (red squares) from Ref. [95]. We presume an uncertainty in the measured data of $0.3\%_{abs}$ in η and of $0.5\%_{abs}$ in FF.*

6.2. Simulation results

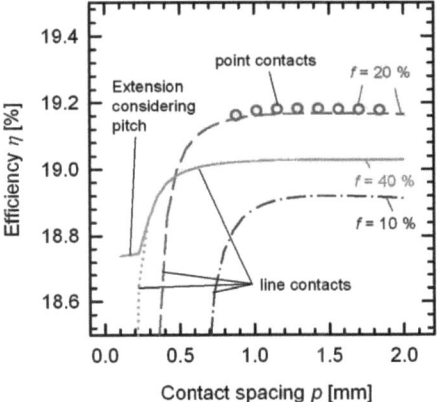

Figure 6.7.: *Simulated efficiency η as a function of the contact spacing p. Line contacts are simulated with a rear metallization fraction of f = 10% (blue dash dotted line), 20% (red dashed line) and 40% with (green solid line) and without the extension in Eqn. 4.21 (green dotted line). Point contacts are simulated with f = 20% (red open circles).*

majority carriers inside the base to the rear contacts with increasing pitch, which increases the base series resistance.

6.2.2. Point or line contacts?

After demonstrating the reliability of the model, we aim at a determination of the optimum rear metallization fraction and study the general difference between point and line contacts. For this purpose we simulate the energy conversion η as a function of the contact spacing p, while keeping the rear metallization fraction constant at $f = 10, 20$ and 40%.

Point contacts are simulated only for pitches $p > 700\mu m$, since an increase in the Al-p^+ layer thickness is expected [48] for smaller contact pitches. As a consequence one would need to consider the impact of p on the contact formation for point contacts with $p < 700\mu m$. We consider the impact of the contact pitch p on contact formation in an extension of our kinetic model (see chapter 4). However, the extension is only valid for line contacts and no corresponding equation for point contacts has been derived yet.

In Fig. 6.7 we determine a maximum in the energy conversion efficiency of $\eta = 19.2\%$ at a metallization fraction of $f_{opt} = 20\%$. This efficiency can be achieved over a wide range of point and line contact spacings between p between $1 - 2mm$. A comparison of the point and line contact data at $f = 20\%$, demonstrates the equality of the two geometries in terms of power conversion efficiency. This is an important result since contact formation at point and line contacts differs significantly according to our findings in chapter 2. However, in terms of contact recombination both point and line contacts are equal in this case.

The optimum metallization of $f_{opt} = 20\%$ for local Aluminum alloyed contacts is relatively high. For other types of contacts, i.e. evaporated contacts usually values below 5% are reported [20, 70]. We will examine the reason for the high f_{opt} later in this section.

As shown in Fig. 6.7 η decreases with decreasing pitch p for metallization fractions of $f = 10$ and 20%. This is a consequence of small Al-p^+ layer thicknesses W_{Al-p^+} occurring in the case of small line contact widths and small f. For $f = 40\%$ we find the same trend with decreasing p (green dotted line).

In chapter 4 we introduced an extension of our kinetic model for contact formation allowing for a description of the impact of contact spacing. Taking into account this extension (green solid line) η stays rather high for small p. Hence, using local Aluminum alloyed contacts allows for the application of small contact pitches in the case of high metallization fractions (in this case $f = 40\%$).

6.2.3. Comparison of local Al alloyed contacts opened with the ps and ns laser

In chapter 5 we found a significant impact of the laser process used for the ablation of the dielectric layer. For this purpose we used in Eqn. 6.2 the parameter S_{min} describing the contact recombination velocity S_{cont} in the case of infinitely thick Al-p^+ layers. This allowed us to consider the impact of the laser process in our simulation model.

We determined S_{min} in Fig. 6.2 for two different laser processes. We obtained $S_{min} = 300 cm/s$ in the case of the ps laser process and $500 cm/s$ for the ns laser process.

We now investigate the impact of S_{min} on solar cell performance in Fig. 6.8. For this purpose we simulate the efficiency as a function of the contact spacing p for the optimum metallization fraction of $f = 20\%$. The higher quality of the ps laser contacts results in an efficiency gain of approximately $0.2\%_{abs}$ compared to the ns laser contacts. This result confirms our understanding that lowest contact recombination is necessary for highly efficient locally contacted solar cells.

6.2.4. Impact of contact resistance

The specific contact resistance $\rho_C = 55 m\Omega cm^2$ reported for local Al alloyed contacts [95] is the reason for relatively low fill factors measured on solar cells. Lower contact resistance values may be achieved by optimizing the Al paste used for screen printing. We analyze

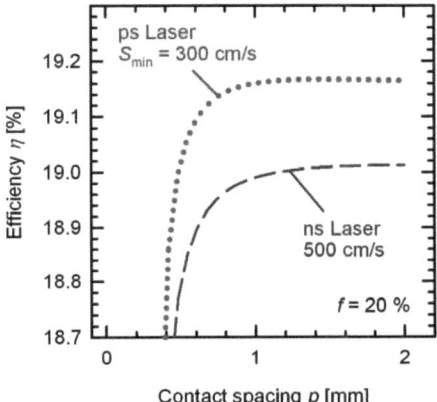

Figure 6.8.: *Demonstrating the impact of S_{min} on solar cell efficiency for the case of a ps laser process ($S_{min,ps} = 300cm/s$, pink dotted line)) and a ns laser process ($S_{min,ns} = 500cm/s$, blue dashed line).*

the impact of lowered specific contact resistances ρ_C of local Al alloyed contacts on the efficiency and the optimum pitch.

The simulations performed for specific contact resistances $\rho_C = 15$, 35 and $55m\Omega cm^2$ are shown in Fig. 6.9. We observe a significant increase in efficiency of $0.3\%_{abs}$ between the two extreme specific contact resistances $\rho_C = 15$ and $55m\Omega cm^2$. Furthermore the optimal pitch for highest efficiencies is dependent on ρ_C. For $\rho_C = 55m\Omega cm^2$ the optimal pitch is $650\mu m$ ($f = 19\%$), whereas for $\rho_C = 15m\Omega cm^2$ we calculate an optimal pitch of $1050\mu m$ ($f = 12\%$). This result indicates a strong limitation of the efficiency due to the high contact resistance of local Aluminum alloyed contacts.

6.3. Conclusion

In this chapter we introduce a comprehensive analytical model for the simulation of screen printed solar cells comprising local Al-alloyed rear contacts for device optimization. For this purpose we consider the impact of the processing conditions on the formation of those contacts and their influence on solar cell performance. Furthermore we include in the simulations an optical model which accounts for the dependence of the rear reflectance on the rear metallization fraction.

Our simulations reproduce literature cell data in terms of the power conversion efficiency

92 Chapter 6. Simulation of silicon solar cells with local Aluminum alloyed base contacts

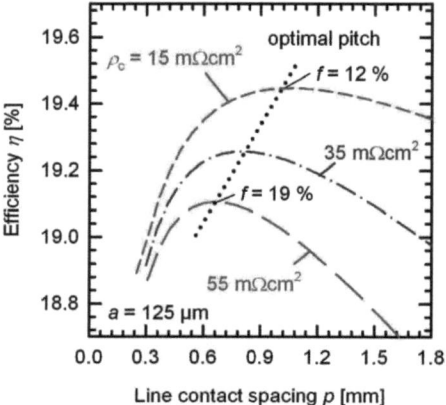

Figure 6.9.: *Power conversion efficiency η simulated as a function of the line contact spacing p. We simulated η using specific contact resistances of $\rho_C = 55$ (red long dashed line), 35 (blue dashed dotted line) and 15 $m\Omega cm^2$ (pink short dashed line). The black dotted line indicates the change in optimal pitch.*

η and FF as a function of the line contact spacing. Our results demonstrate optimum efficiency of point and line contacts at a metallization fraction of $f = 20\%$. This result is in agreement with the contact geometries used for record solar cells at ISFH [9]. Furthermore we observe for line contacts arranged with $f = 40\%$ only a small drop in efficiency of $0.4\%abs$ when decreasing the contact pitch p below $300\mu m$. Comparing the ps laser and ns laser process without KOH etch we observe a $0.2\%abs$ higher efficiency at the ps laser contacts. We identify the high contact resistance as a limiting factor for the fill factor in solar cells with local Al alloyed contacts.

7. Summary

In this work we investigated the properties of local Al alloyed contacts to silicon. The very low recombination of minority charge carriers at such contacts and the simplicity of the manufacturing process makes them ideal candidates for mass production of silicon solar cells. At first we gave a short introduction into the concept of local contacts to silicon solar cells and an explanation of the reduced recombination at local Al alloyed contacts due to the formation of highly aluminum doped (Al-p^+) layers.

We measure contact recombination using a combination of both: a) a spatially resolved charge carrier lifetime measurement using the dynamic Infrared Lifetime Mapping (ILM) technique and b) an analytical model allowing for the separation of recombination in the passivated and contacted region of a sample. For this purpose we apply the Fischer model, which is an approximation of the three dimensional charge carrier transport in a locally contacted sample only. Using a simplification of the Fischer model we identified possible restrictions when measuring the contact recombination velocity S_{cont}. In the case of diffusion limited transport a correct determination of S_{cont} is not possible.

We test our measurement technique for the local contact recombination at local contacts formed by laser ablation of a dielectric stack and subsequent evaporation of Al (LCO). We measure local reverse saturation current densities as low as $J_{0,cont} = 2 \times 10^3 fA/cm^2$ at those contacts. Furthermore laser fired contacts (LFCs) were applied to the same dielectric stack of passivation. We observe no difference in $J_{0,cont}$ between LCO and LFC which is in contrast to the general believe that recombination at LFCs is reduced due to a highly doped layer underneath the contacts. Our results indicate degradation of the passivation stack due to laser treatment in the vicinity of the LCO and LFC.

At local Al alloyed contacts we determine contact reverse saturation current densities as low as $J_{0,cont} = 9 \times 10^2\ fA/cm^2$ on 1.5 Ωcm p-type float-zone (FZ) silicon. This result verifies the assumption of very low recombination at local Al-alloyed contacts. When compared to recombination at LCO and LFC we measure at least one order of magnitude lower J_0 values at samples of a similar resistivity. From scanning electron microscopy (SEM) images we find reduced recombination as a result of more than 1 μm thick Al-p^+ layers. Analyzing the contact formation process as a function of the contact size and layout we show that point contact radii $r > 100\ \mu m$ and line contact widths $a > 80\ \mu m$ are appropriate for lowest contact recombination. In the case of smaller contact widths the Al-p^+ layer is not thick enough to prevent minority charge carrier recombination.

In order to understand the correlation of Al-p^+ layer thickness and local contact geometry, we analyze the contact formation process further. Using quantitative yet simple analytical modeling, the time-dependent silicon concentration c_{Si} in the Al melt is described by elementary differential equations. From these calculations we determine the Al-p^+ layer thickness W_{Al-p^+} and find excellent agreement with experimental data. In contrast to the formation of full area Al-p^+ layers we find that c_{Si} may be smaller than the equilibrium concentration in the phase diagram at the end of the firing process. This is the result of the process dynamics such as the dissolution rate of solid silicon and the transport of silicon in the Al melt. Implementing the impact of the rear contact spacing p on W_{Al-p^+} in our model, we achieve excellent agreement with experimental data.

We compare ps and ns laser ablation processes used to create local Al alloyed contacts. At the contacts processed with ns laser pulses we observe one order of magnitude higer J_0 values. A strong inhomogeniety of the contact formation process and thus W_{Al-p^+} were interpreted as a result of the strong surface roughness after ns laser ablation. Introducing a short KOH etch after laser ablation reduces surface roughness and results in compareable properties of ns and ps laser process in terms of recombination and contact formation.

Finally we apply our findings to simulations of silicon solar cells to indentify the potential of local Al alloyed contacts. For this purpose we extend an optimization tool for the rear contact geometry of solar cells introduced by Wolf et al. with experimentally verified parameterizations of the recombination, series resistance and the rear contact fraction dependent rear reflectance at local Al alloyed contacts. Our study reveals equal performance of point and line rear contact layouts with an optimum metallization fraction of 20 % on 2 Ωcm p-type Czochralski grown silicon.

In contrast to the simplicity in fabrication understanding of local Al alloyed contacts is rather complex. The multiple interdependencies of processing parameters, such as contact geometry, paste composition or firing conditions lead to a broad range of possibilities to manipulate such contacts. This makes it even more important to describe the contact formation of local Al alloyed contacts in a quantitative and basic way to ensure optimum performance of silicon solar cells.

In future research further analysis of the contact formation process may be performed to understand and optimize the properties of local Al-alloyed contacts. Investigating the impact of surface roughness further may help to establish a ns laser ablation process allowing for low contact recombination without KOH etch and to avoid cavity formation. Furthermore it would be interesting to study innovative rear contact geometries for optimum device performance considering the impact of the local contact geometry on the contact formation. Using Al paste containing Si demonstrated interesting results when applied to solar cells. However, a deeper understanding of the underlying mechanisms might be helpful.

A. Comparison of Fischer model and numerical simulations

The Fischer model as introduced in chapter 2 is an analytic approximation of the charge carrier transport in a locally contacted sample [17]. We validate this approximation by performing numerical simulations. For this purpose we employ finite element simulations using the tool Comsol [104].

In our simulation we solve the diffusion equation

$$\nabla^2 n(r) - n(r)/L + G(r)/D = 0 \tag{A.1}$$

in the solar cell base for one specific sample configuration. Here D denotes the minority charge carrier diffusion coefficient. The bulk diffusion length $L = \sqrt{D\tau_{bulk}}$ is related to the bulk lifetime τ_{bulk}. The surface recombination velocities at the front (S_{front}) and at the rear interface (S_{pass} as well as S_{cont}) act as boundary conditions for Eqn. A.1.

Using the solution of the diffusion equation we derive the minority charge carrier distribution $n(r)$ at a given generation rate G. This allows to calculate an effective charge carrier lifetime

$$\tau_{eff} = \frac{\Delta n}{G} \tag{A.2}$$

from the equality of generation rate G and recombination rate $\Delta n / \tau_{eff}$ in steady state conditions. From the effective charge carrier lifetime τ_{eff} we evaluate the effective rear surface recombination velocity $S_{eff,rear}$ with Eqn. 2.5 allowing for a comparison of Fischer's model and our simulation.

In the simulation we consider absorption of the incident light to calculate the depth dependent generation rate of charge carriers. For this purpose we use the absorption law of Lambert-Beer, where the incident light intensity

$$I = I_0 exp^{-\alpha z} \tag{A.3}$$

is decreasing with increasing depth z. We chose a wavelength of $930 nm$ for the incident light corresponding to an absorption length of $\alpha = 210 cm^{-1}$. The chosen wavelength allows to compare our simulation with the dynamic ILM technique. This lifetime measurement

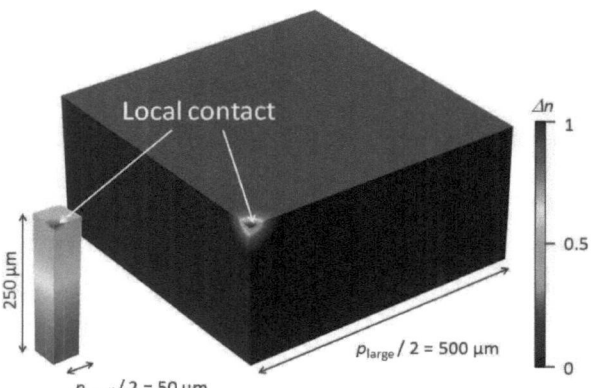

Figure A.1.: *Excess charge carrier distribution Δn in point contacted samples with pitch $p_{small} = 100\mu m$ (left - scale from 10^{14} to $5 \times 10^{14} cm^{-3}$) and $p_{large} = 1mm$ (right - scale from 4×10^{15} to $8 \times 10^{15} cm^{-3}$). Displayed is only the simulated unit cell. Δn is strongly reduced in the area of the local contacts.*

technique is detailed explained in section 2.2.

We simulate a sample with equally spaced point contacts of radius $r = 20\mu m$. By solving the diffusion equation in Eqn. A.1 we obtain the charge carrier distribution depicted in Fig. A.1 for contact pitches of $p_{small} = 100\mu m$ and $p_{large} = 1mm$. An overview of the simulation parameters is given in Tab. A.1.

We observe in Fig. A.1 that the excess charge carrier concentration Δn is strongly reduced in both cases near the local contact. However, the lateral variation of Δn is much more pronounced in the case of the large pitch of $1mm$. Note, that the coloured scale of excess charge carrier density Δn differs strongly between the two geometries. In the case of the geometry with a small contact pitch p_{small} we evaluate a significantly lower $\Delta n \sim 10^{14} cm^{-3}$ compared to the large contact pitch p_{large} with $\Delta n \sim 10^{15} cm^{-3}$.

Using the equality of recombination and generation rate in steady state we calculate the effective charge carrier lifetime

$$\tau_{eff} = \frac{\int \Delta n dV}{\int G dV} \quad (A.4)$$

in the unit cell of volume $\int dV$. In the case of a pitch of $p_{large} = 1mm$ we calculate $\tau_{eff} = 890\mu s$ and in the case of $p_{small} = 100\mu m$ we calculate $\tau_{eff} = 180\mu s$ only.

We calculate the charge carrier distribution for pitches between $p = 100\mu m$ and $5mm$

Table A.1.: *Simulation parameters used in the Comsol simulation of the geometries in Fig. A.1*

Parameter	Value
Point contact radius r	$20\ \mu m$
Sample thickness W	$250\ \mu m$
Diffusion coefficient D	$30\ cm^2/s$
Bulk diffusion length L	$3.3\ mm$
Front surface recombination velocity S_{front}	$10\ cm/s$
Rear surface recombination velocity S_{rear}	$10\ cm/s$
Contact recombination velocity S_{cont}	$10^3\ cm/s$
Photon wavelength λ	$930\ nm$
Illuminating photon flux	$2.78 \times 10^{17}\ cm^{-2}s^{-1}$

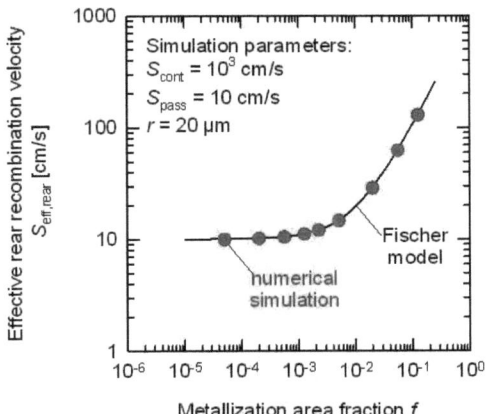

Figure A.2.: *The effective rear surface recombination velocity $S_{eff,rear}$ determined from numerical simulations as a function of the metallization fraction f of point contacts. Equality of the numerical simulation and Fischer model (Eqn. 2.12) is demonstrated.*

and evaluate the effective effective charge carrier lifetime τ_{eff}. From τ_{eff} we determine the rear surface recombination velocity $S_{eff,rear}$ using Eqn. 2.5. We plot the $S_{eff,rear}$ values obtained from our simulation together with the analytic Fischer model as a function of the metallization area fraction f in Fig. A.2. For the Fischer model we used the same input parameters as for the simulation. The equality of numerical simulation and Fischer model validates the approximations made to derive the Fischer model.

B. Other cell simulation results

In chapter 6 we explained a simulation model for Aluminum alloyed contacts and the extensions made in this work. In the following we analyze the impact of the rear reflectance on solar cell level.

B.1. Impact of rear reflectance

For a high short circuit current density J_{sc} of a solar cell good light trapping properties of the device are necessary [17]. This can be achieved with a high rear reflectance allowing for low energy photons to pass the solar cell twice or more and hence to increase the probability of photon absorption [17, 105].

Usually the rear reflectance at local metal contacts is neglected due to a low optimum metallization fraction $f_{opt} < 5\%$ [20, 70]. However, in the case of local Aluminum alloyed contacts a relatively high metallization fraction $f_{opt} = 20\%$ is necessary for optimum device performance. Hence, we consider the impact of the reflectance at the local rear contacts in our simulation.

Figure B.1 shows the impact of the rear reflectance R on the simulated short circuit current density J_{sc} in the case of local Al alloyed contacts. In one case (red solid line in Fig. B.1) we perform simulations considering the rear metallization fraction dependent reflectance by means of Eqn. 6.6. In the other case (dashed black line in Fig. B.1) we assume a constant $J_{gen,hom} = J_{gen,hom}(f = 0)$ equal to the case when no contacts would have been applied. For both cases Fig. B.1 shows the energy conversion efficiency as a function of the line contact spacing for a line width of $a = 125 \mu m$. We observe in Fig. B.1 a difference in J_{sc} as high as $0.75 mA/cm^2$. It decreases with increasing pitch p.

We simulate the power conversion efficiency η in Fig. B.2 for the two cases in Fig. B.1. As expected, we observe higher efficiency values for the constant $J_{gen,hom} = J_{gen,hom}(f = 0)$. Taking into account the impact of the rear metallization fraction on $J_{gen,hom}$ by the use of Eqn. 6.6 we observe a decrease in efficiency of around $0.07\%_{abs}$.

Note, that this is the error which is made using the original model by Wolf et al. [20] without our extension. The decrease in the efficiency is more pronounced for a small pitch p, where the metallized area fraction is high. Furthermore we observe a shift in the optimum pitch of around $60 \mu m$ towards larger contact spacings.

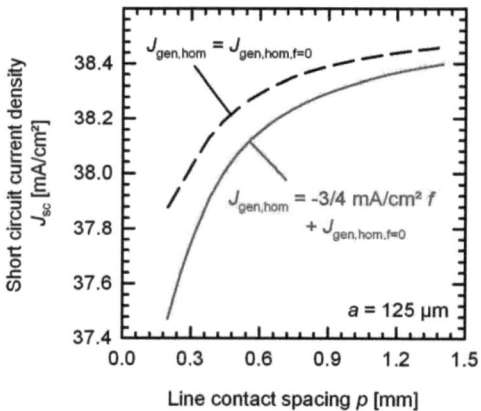

Figure B.1.: *Calculated short circuit current density J_{sc} simulated as a function of the line contact spacing using a line width $a = 125\mu m$. For the simulation we consider a rear metallization fraction dependent reflectance (red solid line) and in the other case a constant $J_{gen,hom} = J_{gen,hom}(f = 0)$ equal to the case when no contacts would have been applied (dashed black line).*

B.1. Impact of rear reflectance

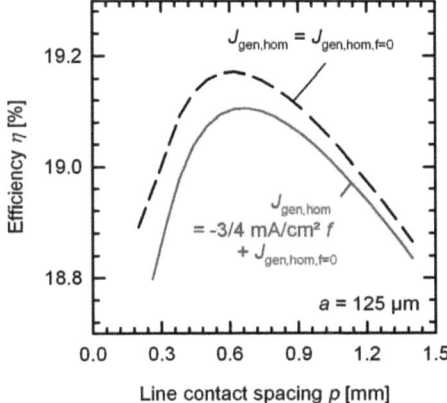

Figure B.2.: *The efficiency η simulated as a function of the line contact spacing p using a line width $a = 125\mu m$. The case of employing Eqn. 6.6 (red solid line) is compared to the case of rear metallization fraction independent optics, which equal the case when no contacts would have been applied.*

References

[1] J. Mandelkorn and J. Lamneck, Simplified fabrication of back surface electric field silicon cells and novel characterisation of such cells, *Proceedings of the 9th IEEE Photovoltaic Specialists Conference, Silver Springs, MD, USA*, 66–71 (1972).

[2] B. Hallam, S. Wenham, A. Sugianto, L Mai, C. Chong, M. Edwards, D. Jordan, and P. Fath, Record large-area p-type cz production cell efficiency of 19.3% based on ldse technology, *IEEE Journal of Photovoltaics* **1**(1), 43–48 (2011).

[3] R. Swanson, S. Beckwith, R. Crane, W. Eades, Y. Kwark, R. Sinton, and S. Swirhun, Point-contact silicon solar cells, *IEEE Trans. Electron Devices* **31**(5), 661–664 (1984).

[4] A. Blakers, A. Wang, A. Milne, J. Zhao, and M. Green, 22.8% efficient silicon solar cell, *Appl. Phys. Lett.* **55**, 1363–1365 (1989).

[5] J. Zhao, A. Wang, and M. Green, 24% efficient PERL structure silicon solar cells, *Proceedings of the 21st IEEE Photovoltaic Specialists Conference, Kissimmee, FL, USA*, pp. 333–335 (1990).

[6] G. Agostinelli, P. Choulat, H. Dekkers, Y. Ma, and G. Beaucarne, Silicon solar cells on ultra-thin substrates for large scale production, *Proceedings of the 21st European Photovoltaic Solar Energy Conference, Dresden, Germany*, pp. 601–604 (2006).

[7] Vichai Meemongkolkiat, Kenta Nakayashiki, Dong Kim, Steve Kim, Aziz Shaikh, Armin Kuebelbeck, Werner Stockum, and Ajeet Rohatgi, Investigation of modified screen-printing Al pastes for local back surface field formation, *Proceedings of the 32nd IEEE Photovoltaic Specialists Conference, Waikoloa, HI, USA*, pp. 1457–1470 (2006).

[8] J. Lai, A. Upadhyaya, S. Ramanathan, A. Das, K. Tate, V. Upadhyaya, A. Kapoor, C. Chen, and A. Rohatgi, High efficiency large-area rear passivated silicon solar cells with local al-bsf and screen-printed contacts, *IEEE Journal of Photovoltaics* **1**(1), 16–21 (2011).

[9] B. Veith, T. Dullweber, M. Siebert, C. Kranz, F. Werner, N. Harder, J. Schmidt, B. Roos, T. Dippell, and R. Brendel, Comparison of icp-alox and ald-al2o3 layers for the rear surface passivation of c-si solar cells, *Energy Procedia* **27**, 379–384 (2012).

[10] A. Lachowicz, K. Ramspeck, P. Roth, M. Manole, H. Blanke, W. Hefner, E. Brouwer, B. Schum, and A. Metz, Nox free solution for emitter etch-back, *Proceedings of the 27th European Photovoltaic Solar Energy Conference, Frankfurt, Germany*, 1846–1850 (2012).

[11] R. Bock, J. Schmidt, R. Brendel, H. Schuhmann, and M. Seibt, Electron microscopy analysis of crystalline silicon islands formed on screen-printed aluminum-doped p-type silicon surfaces, *J. Appl. Phys.* **104**, 043701 (2008).

[12] A Rohatgi and P. Rai Choudhury, Design, farbication, and analysis of 17-18 percent efficient surface-passivated silicon solar cells, *IEEE Trans. Electron Devices* **31**(5), 596–601 (1984).

[13] B. Fischer, *Metallisation of Silicon Solar cells*, Diploma thesis, Australian national university Canberra and Fachhochschule Ravensburg-Weingarten (1994).

[14] A. Uruena, J. John, G. Beaucarne, P. Choulat, P. Eyben, G. Agostinelli, E. Van Kerschaver, J. Poortmans, and R. Mertens, Local al-alloyed contacts for next generation si solar cells, *Proceedings of the 24th European Photovoltaic Solar Energy Conference, Hamburg, Germany*, 1483–1486 (2009).

[15] F. Grasso, L. Gautero, J. Rentsch, R. Preu, and R. Lanzafame, Characterisation of local al-bsf formation for perc solar cell structures, *Proceedings of the 25th European Photovoltaic Solar Energy Conference, Valencia, Spain* (2010).

[16] K. Ramspeck, S. Reissenweber, J. Schmidt, K. Bothe, and R. Brendel, Dynamic carrier lifetime imaging of silicon wafers using an infrared-camera-based approach, *Appl. Phys. Lett.* **93**, 102104 (2008).

[17] B. Fischer, *Loss analysis of crystalline silicon solar cells using photoconductance and quantum efficiency measurements*, PhD thesis, University of Konstanz (2003).

[18] W. Pfleging, A. Ludwig, K. Seemann, R. Preu, H. Mäckel, and S. Glunz, Laser micromachining for applications in thin film technology, *Applied Surface Science* **154-155**, 633–639 (2000).

[19] E. Schneiderlöchner, R. Preu, R. Lüdemann, and S. Glunz, Laser-fired rear contacts for crystalline silicon solar cells, *Prog. Photovolt: Res. Appl.* **10**, 29–34 (2002).

[20] A. Wolf, D. Biro, J. Nekarda, S. Stumpp, A. Kimmerle, S. Mack, and R. Preu, Comprehensive analytical model for locally contacted rear surface passivated solar cells, *Journal of Applied Physics* **108** (2010).

[21] S. Sterk, J. Knobloch, and W. Wettling, Optimization of the rear contact pattern of high-efficiency silicon solar cells with and without local back surface field, *Prog. Photovolt: Res. Appl.* **2**, 19–26 (1994).

[22] A. Aberle, G. Heiser, and M. Green, Two-dimensional numerical optimization study of the rear contact geometry of high-efficiency silicon solar cells, *J. Appl. Phys.* **75**, 5391–5405 (1994).

[23] U. Rau, Three dimensional simulation of the electrical transport in high efficiency solar cells, *Proceedings of the 12th European Photovoltaic Solar Energy Conference, Amsterdam, Netherlands*, pp. 1350–1353 (1994).

[24] K. Catchpole and A. Blakers, Modelling the PERC structure for industrial quality silicon, *Sol. Energy Mater. Sol. Cells* **73**, 189–202 (2002).

[25] D. Kray and K. McIntosh, Analysis of ultrathin high-efficiency silicon solar cells, *Phys. Status Solidi A* **206**, 1647–1654 (2009).

[26] Electrical Properties of Silicon, *http://www.ioffe.rssi.ru/SVA/NSM/Semicond/Si/electric.html*.

[27] J. Schmidt, *Untersuchungen zur Ladungsträgerrekombination an der Oberfläche und im Volumen von kristallinen Silicum-Solarzellen*, PhD thesis, University of Hannover (1998).

[28] H. Plagwitz, *Surface passivation of crystalline silicon solar cells by amorphous silicon films*, PhD thesis, Universität Hannover (2007).

[29] B. Godfrey and M. Green, High efficiency silicon minmis solar cells - design and experimental results, *IEEE Trans. Electron Devices* **27**(4), 737–745 (1980).

[30] A. Aberle, B. Kuhlmann, R. Meyer, A. Hübner, C. Hampe, and R. Hezel, Comparison of p-n junction and inversion layer silicon solar cells by means of experiment and simulation, *Prog. Photovolt: Res. Appl.* **4**, 193–204 (1996).

[31] D. Zielke, J. Petermann, F. Werner, B. Veith, R. Brendel, and J. Schmidt, Contact passivation in silicon solar cells using atomic layer deposited aluminum oxide layers, *Phys. Status Solidi RRL* **5**, 298–300 (2011).

[32] M. Taguchi, K. Kawamoto, S. Tsuge, T. Baba, H. Sakata, M. Morizane, K. Uchihashi, N. Nakamura, S. Kiyama, and O. Oota, Hit cells - high efficiency crystalline silicon cells with novel structure, *Prog* **8**, 503–513 (2000).

[33] M. Godlewski, C. Baraona, and H. Brandhorst, Low-high junction theory applied to solar cells, *Proceedings of the 10th IEEE Photovoltaic Specialists Conference, Palo Alto, CA, USA*, pp. 40–49 (1973).

[34] S. Glunz, E. Schneiderlöchner, D. Kray, A. Grohe, M. Hermle, H. Kampwerth, R. Preu, and G. Willeke, Laser-fired contact silicon solar cells on p- and n-substrates, *Proceedings of the 19th European Photovoltaic Solar Energy Conference, Paris, France*, pp. 408–411 (2004).

[35] V. Naumann, C. Hagendorf, M. Werner, B. Henke, C. Schmidt, J. Nekarda, and J. Bagdahn, Local electronic properties and microstructure of individual laser fired contacts, *Proceedings of the 24th European Photovoltaic Solar Energy Conference, Hamburg, Germany* (2009).

[36] R. Ferre, R. Gogolin, J. Müller, N. Harder, and R. Brendel, Laser transfer doping for contacting n-type crystalline si solar cells, *Phys. Status Solidi A* **208**(8), 1964–1966 (2011).

[37] A. Fell, E. Franklin, D. Walter, D. Suh, and K. Weber, Laser doping from al2o3 layers, *Proceedings of the 27th European Photovoltaic Solar Energy Conference, Frankfurt, Germany*, 706–708 (2012).

[38] R. Bock, J. Schmidt, and R. Brendel, n-type silicon solar cells with surface-passivated screen-printed aluminium-alloyed rear emitter, *Phys. Status Solidi RRL* **2**, 248–250 (2008).

[39] R. Woehl, J. Krause, F. Granek, and D. Biro, 19.7% efficient all screen printed back contact back junction silicon solar cell with aluminum alloyed emitter, *IEEE Electron Device Letters* **32**(3), 345–347 (2011).

[40] S. Narashima, A. Rohatgi, and A. Weeber, An optimized rapid aluminum back surface field technique for silicon solar cells, *IEEE Trans. Electron Devices* **46**(7), 1363–1999 (1999).

[41] S. Gatz, H. Hannebauer, R. Hesse, F. Werner, A. Schmidt, T. Dullweber, J. Schmidt, K. Bothe, and R. Brendel, 19.4%-efficient large-area fully screen-printed silicon solar cells, *Phys. Status Solidi RRL* **5**, 147–149 (2011).

[42] E. Cornagliotti, A. Uruena, J. Horzel, J. John, L. Tous, D. Hendrickx, V. Prajapati, S. Singh, R. Hoyer, F. Delahaye, K. Weise, D. Queisser, H. Nussbaumer, and J. Poortmanss, How much rear side polishing is required? a study on the impact of rear side polishing in perc solar cells, *Proceedings of the 27th European Photovoltaic Solar Energy Conference, Frankfurt, Germany* (2012).

[43] C. Kranz, S. Wyczanowski, S. Dorn, K. Weise, C. Klein, K. Bothe, T. Dullweber, and R. Brendel, Impact of the rear surface roughness on industrial-type perc solar cells, *Proceedings of the 27th European Photovoltaic Solar Energy Conference, Frankfurt, Germany* (2012).

[44] J. Müller, K. Bothe, S. Gatz, H. Plagwitz, G. Schubert, and R. Brendel, Recombination at local aluminum-alloyed silicon solar cell base contacts by dynamic infrared lifetime mapping, *Energy Procedia* **8**, 337–342 (2011).

[45] J. Müller, K. Bothe, S. Gatz, H. Plagwitz, G. Schubert, and R. Brendel, Contact formation and recombination at screen-printed local aluminum-alloyed silicon solar cell base contacts, *IEEE Trans. Electron Devices* **58**, 3239–3245 (2011).

[46] J. Müller, K. Bothe, S. Gatz, and R. Brendel, Modeling the formation of local highly aluminum doped silicon regions by rapid thermal annealing of screen-printed aluminum, *Phys. Status Solidi RRL* **1-3** (2012).

[47] E. Urrejola, K. Peter, H. Plagwitz, and G. Schubert, Distribution of silicon in the aluminum matrix for rear passivated solar cells, *Energy Procedia* **8**, 331–336 (2011).

[48] M. Rauer, C. Schmiga, R. Woehl, K. Rühle, M. Hermle, M. Hörteis, D. Biro, and S. Glunz, Investigation of aluminum alloyed local contacts for rear surface passivated si solar cells, *IEEE J. Photovoltaics* **1**(1), 22–28 (2011).

[49] M. Bähr, G. Heinrich, O. Doll, I. Köhler, C. Maier, and A. Lawerenz, Differences of rearc-contact formation between laser ablation and etching paste for perc solar cells, *Proceedings of the 26th European Photovoltaic Solar Energy Conference, Hamburg, Germany* (2011).

[50] J. Murray and A. McAlister, The aluminum-silicon system, *Bull. Alloy Phase Diagrams* **5**, 74 (1984).

[51] J. Song, S. Park, S. Kwon, S. Kim, H. Kim, S. Tark, S. Yoon, and D. Kim, A study on the aluminum fire-through to a a-sinx:h thin film for crystalline solar cells, *Current Applied Physics* **12**, 313–318 (2012).

[52] I. Cesar, E. Granneman, P. Vermont, H. Kathri, H. Kerp, A. Shaikh, P. Manshanden, A. Mewe, I. Romijn, and A. Weeber, Industrial application of uncapped al2o3 and firing-through al-bsf in open rear passivated solar cells, *Proceedings of the 37th IEEE Photovoltaic Specialists Conference, Seattle, WA, USA* (2011).

[53] S. Ramanathan, V. Meemongkolkiat, A. Das, A. Rohatgi, and I. Koehler, 20% efficient screen printed lbsf cell fabricated using uv laser for rear dielectric removal,

Proceedings of the 35th IEEE Photovoltaic Specialists Conference, Hawaii, USA, 678–682 (2010).

[54] E. Urrejola, K. Peter, H. Plagwitz, and G. Schubert, Silicon diffusion in aluminum for rear passivated solar cells, *Appl. Phys. Lett.* **98** (2011).

[55] D. Chen, Z. Liang, Y. Yang, H. Shen, and Y. Liu, Structure simulation of screen printed local back surface field for rear passivated silicon solar cells, *Proceedings of the 38th IEEE Photovoltaic Specialists Conference, Austin, Tx, USA* (2012).

[56] K. Dressler, S. Dauwe, T. Droste, J. Rossa, B. Meidel, K. Schünemann, K. Ramspeck, Y. Gassenbauer, and A. Metz, Characterisation of rear local contacts including bsf formation using raman and scanning acoustic microscopy, *Proceedings of the 27th European Photovoltaic Solar Energy Conference, Frankfurt, Germany* (2012).

[57] D. Chen, Y. Yang, Z. Li, Z. Liang, Z. Feng, P. Verlinden, and H. Shen, Analysis of morphologies and distribution of al-doped local back surface field for screen printed i-perc solar cell, *Proceedings of the 27th European Photovoltaic Solar Energy Conference, Frankfurt, Germany* (2012).

[58] J. Müller, K. Bothe, S. Gatz, F. Haase, C. Mader, and R. Brendel, Recombination at laser-processed local base contacts by dynamic infrared lifetime mapping, *J. Appl. Phys.* **108**, 124513 (2010).

[59] A.B. Sproul, Dimensionless solution of the equation describing the effect of surface recombination on carrier decay in semiconductors, *J. Appl. Phys.* **76**(5), 2851–2854 (1994).

[60] C. Mader, *In-line high-rate evaporation of aluminum for the metallization of silicon solar cells*, PhD thesis, University of Hanover (2012).

[61] K. Ramspeck, K. Bothe, J. Schmidt, and R. Brendel, Combined dynamic and steady-state infrared camera based carrier lifetime imaging of silicon wafers, *J. Appl. Phys.* **106**, 114506 (2009).

[62] K. Ramspeck, *Characterization techniques for silicon solar cells and material using an Infrared-camera based approach*, PhD thesis, University of Hanover (2009).

[63] M. Bail, J. Kentsch, R. Brendel, and M. Schulz, Lifetime mapping of Si wafers by an infrared camera, *Proceedings of the 28th IEEE Photovoltaic Specialists Conference, Anchorage, AK, USA*, pp. 99–103 (2000).

[64] A. Cuevas, Geometrical analysis of solar cells with partial rear contacts, *IEEE J. Photovoltaics* **2**(4), 485–493 (2012).

[65] A. Cuevas, Physical model of back line-contact front-junction solar cells, *J. Appl. Phys.* **113** (2013).

[66] R. Cox and H. Strack, Ohmic contacts for GaAs devices, *Solid State Electron.* **10**, 1213–1218 (1967).

[67] B. Gelmont, M. Shur, and R. Mattauch, Disk and stripe capacitances, *Solid State Electron.* **38**, 731–734 (1995).

[68] H. Plagwitz and R. Brendel, Analytical model for the diode saturation current of point-contacted solar cells, *Prog. Photovolt: Res. Appl.* **14**, 1–12 (2006).

[69] P. Saint-Cast, M. Rüdiger, A. Wolf, M. Hofmann, J. Rentsch, and R. Preu, Advanced analytical model for the effective recombination velocity of locally contacted surfaces, *J. A* **108** (2010).

[70] C. Mader, J. Müller, S. Gatz, T. Dullweber, and R. Brendel, Rear side point-contacts by in-line thermal evaporation of aluminum, *Proceedings of the 35th IEEE Photovoltaic Specialists Conference, Honolulu, HI, USA*, pp. 1446–1449 (2010).

[71] M. Kerr and A. Cuevas, General parameterization of Auger recombination in crystalline silicon, *J. Appl. Phys.* **91**, 2473–2480 (2002).

[72] M. Schöfthaler, U. Rau, and J. Werner, Direct observation of a scaling effect on effective minority carrier lifetimes, *J. Appl. Phys.* **76**, 4168–4172 (1994).

[73] S. Hermann, N. Harder, and R. Bren, Picosecond laser ablation of sio2 layers on silicon substrates, *Appl. Phys. A* (2009).

[74] F. Haase, T. Neubert, R. Horbelt, B. Terheiden, K. Bothe, and R. Brendel, Local aluminum-silicon contacts by layer selective laser ablation, *Sol. Energy Mater. Sol. Cells* **95**(9), 2698–2700 (2011).

[75] M. Schöfthaler and R. Brendel, Sensitivity and transient response of microwave reflection measurements, *J. Appl. Phys.* **77**, 3162–3173 (1995).

[76] D. Kray and S. Glunz, Investigation of laser-fired rear-side recombination properties using an analytical model, *Prog. Photovolt: Res. Appl.* **14**, 195–201 (2006).

[77] J. Nekarda, D. Reinwand, A. Grohe, P. Hartmann, R. Preu, R. Trassl, and S. Wiede, Industrial PVD metallization for high efficiency silicon solar cells, *Proceedings of the 34th IEEE Photovoltaic Specialists Conference, Philadelphia, PA, USA*, pp. 892–896 (2009).

[78] K. Mangersnes and S. Foss, Laser ablation of pecvd oxide for structuring of back-junction interdigitated silicon solar cells, *Proceedings of the 24th European Photovoltaic Solar Energy Conference, Hamburg, Germany*, 2001–2003 (2009).

[79] O. Schultz, M. Hofmann, S. Glunz, and G. Willeke, Silicon oxide - silicon nitride stack system for 20% efficient silicon solar cells, *Proceedings of the 31st IEEE Photovoltaic Specialists Conference, Lake Buena Vista, FL, USA*, 872–876 (2005).

[80] S. Gatz, T. Dullweber, J. Müller, and R. Brendel, Analysis and optimization of the bulk and rear recombination of screen-printed perc solar cells, *Energy Procedia* (2012).

[81] P. Altermatt, S. Steingrube, Y. Yang, C. Sprodowski, T. Dezhdar, S. Koc, B. Veith, S. Hermann, R. Bock, K. Bothe, J. Schmidt, and R. Brendel, Highly predictive modelling of entire si solar cells for industrial applications, *Proceedings of the 24th European Photovoltaic Solar Energy Conference, Hamburg, Germany* (2009).

[82] J. del Alamo, J. Eguren, and A. Luque, Operating limits of al alloyed high low junctions for bsf solar cells, *Solid-State Electronics* **24**, 415–420 (1981).

[83] C. Sealy, M. Castell, and P. Wilshaw, Mechanism for secondary electron dopant contrast in the SEM, *J. Electron Microsc.* **49**, 311–321 (2000).

[84] D. Clugston and P. Basore, Modelling free-carrier absorption in solar cells, *Prog. Photovolt: Res. Appl.* **5**, 229–236 (1997).

[85] J. Schmidt, N. Thiemann, R. Bock, and R. Brendel, Recombination lifetimes in highly aluminum doped silicon, *J. Appl. Phys.* **106** (2009).

[86] M. Rüdiger, M. Rauer, C. Schmiga, and M. Hermle, Effect of incomplete ionization for the description of highly aluminum-doped silicon, *J. Appl. Phys.* **110** (2011).

[87] F. Huster, Aluminium back surface field: bow investigation and elimination, *Proceedings of the 20th European Photovoltaic Solar Energy Conference, Barcelona, Spain*, pp. 635–638 (2005).

[88] K. Wijekoon, H. Kathri, D. Tanner, L. Zhang, A. Shaikh, and H. Ponnekanti, Rear passivated high efficiency solar cells: Optimization of aluminum allalloy in local contacts by modifying paste formulation, *Proceedings of the 27th European Photovoltaic Solar Energy Conference, Frankfurt, Germany*, 608–613 (2012).

[89] T. Lauermann, B. Fröhlich, G. Hahn, and B. Terheiden, Diffusion-based model of local al back surface field formation for industrial passivated emitter and rear cell solar cells, *Prog. Photovolt: Res. Appl.* (2013).

[90] B. Lenkeit, S. Steckemetz, F. Artuso, and R. Hezel, Excellent thermal stability of remote plasma-enhanced chemical vapour deposited silicon nitride films for the rear of screen-printed bifacial silicon solar cells, *Sol. Energy Mater. Sol. Cells* **65**, 317–323 (2001).

[91] S. Ramanathan, V. Meemongkolkiat, A. Das, A Rohatgi, and I. Koehler, Fabrication of 20% efficient cells using spin-on based simultaneous diffusion and dielectric anneal, *Proceedings of the 34th IEEE Photovoltaic Specialists Conference, Philadelphia, PA, USA*, 2150–2153 (2009).

[92] P. Engelhart, S. Hermann, T. Neubert, H. Plagwitz, R. Grischke, R. Meyer, U. Klug, A. Schoonderbeek, U. Stute, and R. Brendel, Laser ablation of SiO2 for locally contacted si solar cells with ultra-short pulses, *Prog. Photovolt: Res. Appl.* **15**, 521–527 (2007).

[93] S. Eidelloth, T. Neubert, T. Brendemühl, S. Hermann, P. Giesel, and R. Brendel, High speed laser structuring of crystalline silicon solar cells, *Proceedings of the 34th IEEE Photovoltaic Specialists Conference, Philadelphia, PA, USA* (2009).

[94] Z. Du, N. Palina, J. Chen, F. Lin, M. Hong, and B. Hoex, Impact of koh etching on laser damage removal and contact formation for al local back surface field silicon wafer solar cells, *Proceedings of the 27th European Photovoltaic Solar Energy Conference, Frankfurt, Germany*, 1230–1233 (2012).

[95] S. Gatz, T. Dullweber, and R. Brendel, Contact resistance of local rear side contacts of screen-printed silicon PERC solar cells with efficiencies up to 19.4%, *Proceedings of the 36th IEEE Photovoltaic Specialists Conference, Seattle, WA, USA*, pp. 3658–3664 (2011).

[96] A. Li, W. Chen, L. Wang, H. LiuH. Liu. Lai, C. Huang, Y. Lin, T. Fang, and W. Chen, Analysis for interaction of laser ablation metallization on the screen printed rear side local contact, *Proceedings of the 27th European Photovoltaic Solar Energy Conference, Frankfurt, Germany* (2012).

[97] J. Müller, S. Gatz, K. Bothe, and R. Brendel, Optimizing the geometry of local aluminum alloyed contacts to fully screen-printed silicon solar cells, *Proceedings of the 38th IEEE Photovoltaic Specialists Conference, Austin, Tx, USA* (2012).

[98] S. Steingrube, H. Wagner, H. Hannebauer, S. Gatz, R. Chen, S. Dunham, T. Dullweber, P. Altermatt, and R. Brendel, Loss analysis and improvements of industrially fabricated cz-si solar cells by means of process and device simulations, *Energy Procedia* **8**, 263–268 (2011).

[99] J. Schmidt, B. Lim, D. Walter, K. Bothe, S. Gatz, T. Dullweber, and P. Altermatt, Impurity related limitations of next generation industrial silicon solar cells, *IEEE J. Photovoltaics* **3**(1), 114–118 (2013).

[100] E. Urrejola, K. Peter, H. Plagwitz, and G. Schubert, Al-Si alloy formation in narrow p-type Si contact areas for rear passivated solar cells, *J. Appl. Phys.* **107**, 124516 (2010).

[101] R. Brendel, M. Hirsch, R. Plieninger, and J. Werner, Quantum efficiency analysis of thin layer silicon solar cells with back surface fields and optical confinement, *IEEE Trans. Electron Devices* **43**(7), 1104–1113 (1996).

[102] K. Bothe, R. Sinton, and J. Schmidt, Fundamental boron-oxygen-related carrier lifetime limit in mono- and multicrystalline silicon, *Prog. Photovolt: Res. Appl.* **13**, 287–296 (2005).

[103] B. Lim, PhD thesis, University of Hanover (2012).

[104] http://www.comsol.com, *Comsol*.

[105] E. Yablonovitch, Statistical ray optics, *J. Opt. Soc. Am.* **72**(7), 899–907 (1982).

List of Publications

Refereed journal papers as first author

1. J. MÜLLER, K. BOTHE, S. GATZ, F. HAASE, C. MADER, AND R. BRENDEL, "Recombination at laser-processed local base contacts by dynamic infrared lifetime mapping", *J. Appl. Phys.* **108**, 124513 (2010).

2. J. MÜLLER, K. BOTHE, S. GATZ, H. PLAGWITZ, G. SCHUBERT, AND R. BRENDEL, "Contact Formation and Recombination at Screen-Printed Local Aluminum-Alloyed Silicon Solar Cell Base Contacts", *IEEE Trans. Electr. Dev.* **58**, 10 (2011).

3. J. MÜLLER, K. BOTHE, S. GATZ, AND R. BRENDEL, "Modeling the formation of local highly aluminum-doped silicon regions by rapid thermal annealing of screen-printed aluminum", *Phys. Status Solidi RRL* **1-3** (2012).

4. J. MÜLLER, K. BOTHE, S. HERLUFSEN, T. OHRDES, AND R. BRENDEL, "Reverse Saturation Current Density Imaging of Highly Doped Regions in Silicon Employing Photoluminescence Measurements", *IEEE J. Photovoltaics* **2**, 4 (2012).

5. J. MÜLLER, K. BOTHE, S. HERLUFSEN, H. HANNEBAUER, R. FERRÉ, AND R. BRENDEL, "Reverse saturation current density imaging of highly doped regions in silicon: A photoluminescence approach", *Sol. Energy Mat. Solar Cells* **106** (2012).

6. J. MÜLLER, H. HANNEBAUER, C. MADER, F. HAASE AND K. BOTHE, "Dynamic infrared lifetime mapping for the measurement of the saturation current density of highly doped regions in silicon", *IEEE J. Photovoltaics* **reviewed** (2013).

Refereed journal papers as coauthor

1. S. Herlufsen, K. Ramspeck, D. Hinken, A. Schmidt, J. Müller, K. Bothe, J. Schmidt, R. Brendel, "Dynamic photoluminescence lifetime imaging for the characterisation of silicon wafers", *Phys. Status Solidi RRL* **1-3**, (2010).

2. R. Ferré, R. Gogolin, J. Müller, N. P. Harder, and R. Brendel, "Laser transfer doping for contacting n-type crystalline Si solar cells", *Phys. Status Solidi A* **1-3** (2011).

3. O. Breitenstein, J. Bauer, K. Bothe, D. Hinken, J. Müller, W. Kwapil, M. C. Schubert, and W. Warta, "Can Luminescence Imaging Replace Lock-in Thermography on Solar Cells?", *IEEE J. Photovoltaics* **1**, 2 (2011).

4. C. Mader, J. Müller, S. Eidelloth, and R. Brendel, "Local rear contacts to silicon solar cells by in-line high-rate evaporation of aluminum", *Sol. Energy Mat. Solar Cells* **107** (2012).

Papers presented at international conferences as first author

1. J. Müller, K. Bothe, S. Gatz, H. Plagwitz, G. Schubert, and R. Brendel, "Recombination at local aluminum-alloyed silicon solar cell base contacts by dynamic infrared lifetime mapping", *Proceedings of the 1st SiliconPV, Freiburg, Germany*, Energy Procedia **8**, pp. 337-342 (2011).

2. J. Müller, S. Gatz, K. Bothe and R. Brendel, "Optimizing the geometry of local aluminum-alloyed contacts to fully screen-printed silicon solar cells", *Proc. 38th Photovoltaic Specialist Conference, Austin, TX, USA*, pp. 2223 - 2228 , (2012).

Papers presented at international conferences as coauthor

1. C. Mader, J. Müller, S. Gatz, T. Dullweber, and R. Brendel, "Rear side point-contacts by in-line thermal evaporation of aluminum", *Proceedings of the 35th IEEE Photovoltaic Specialists Conference, Honolulu, HI, USA*, pp. 1446-1449, (2010).

2. S. HERLUFSEN, K. RAMSPECK, D. HINKEN, J. MÜLLER, K. BOTHE, J. SCHMIDT, AND R. BRENDEL, "Advances in Camera-Based Photoluminescence Lifetime Imaging", *Proceedings of the 25th European Photovoltaic Solar Energy Conference, Valencia, Spain* (2010).

3. C. MADER, R. BOCK, J. MÜLLER, J. SCHMIDT, AND R. BRENDEL, "Formation of locally aluminum-doped p-type silicon regions by in-line high-rate evaporation", *Proceedings of the 1st SiliconPV, Freiburg, Germany*, Energy Procedia **8**, pp. 521-526 (2011).

4. S. GATZ, K. BOTHE, J. MÜLLER, T. DULLWEBER, AND R. BRENDEL, "Analysis of local Al-doped back surface fields for high efficiency screen-printed solar cells", *Proceedings of the 1st SiliconPV, Freiburg, Germany*, Energy Procedia **8**, pp. 318-323 (2011).

5. R. FERRE, R. GOGOLIN, J. MÜLLER, N. HARDER, M. KESSLER, C. MADER, P. GIESEL, T. NEUBERT, R. BRENDEL, "Laser transfer doping using amorphous silicon", *Proceedings of the 26th European Photovoltaic Solar Energy Conference, Hamburg, Germany*, pp. 879-883, (2011).

6. S. GATZ, J. MÜLLER, T. DULLWEBER, AND R. BRENDEL, "Analysis and Optimization of the Bulk and Rear Recombination of Screen-Printed PERC Solar Cells", *Proceedings of the 2nd SiliconPV, Leuven, Belgium*, Energy Procedia **27**, pp. 95-102 (2012).

7. H. HANNEBAUER, M. SOMMERFELD, J. MÜLLER, T. DULLWEBER, AND R. BRENDEL, "Analysis of the emitter saturation current density of industrial type silver screen-printed front contacts", *Proceedings of the 27th European Photovoltaic Solar Energy Conference, Frankfurt, Germany* (2012).

8. B. FISCHER, J. MÜLLER AND P. ALTERMATT, "A simple emitter model for quantum efficiency curves and extracting the emitter saturation current", *Proceedings of the 28th European Photovoltaic Solar Energy Conference, Paris, France* (2013).

i want morebooks!

Buy your books fast and straightforward online - at one of world's fastest growing online book stores! Environmentally sound due to Print-on-Demand technologies.

Buy your books online at
www.get-morebooks.com

Kaufen Sie Ihre Bücher schnell und unkompliziert online – auf einer der am schnellsten wachsenden Buchhandelsplattformen weltweit! Dank Print-On-Demand umwelt- und ressourcenschonend produziert.

Bücher schneller online kaufen
www.morebooks.de

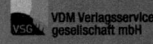 VDM Verlagsservicegesellschaft mbH
Heinrich-Böcking-Str. 6-8 Telefon: +49 681 3720 174 info@vdm-vsg.de
D - 66121 Saarbrücken   Telefax: +49 681 3720 1749 www.vdm-vsg.de

Printed by Books on Demand GmbH, Norderstedt / Germany